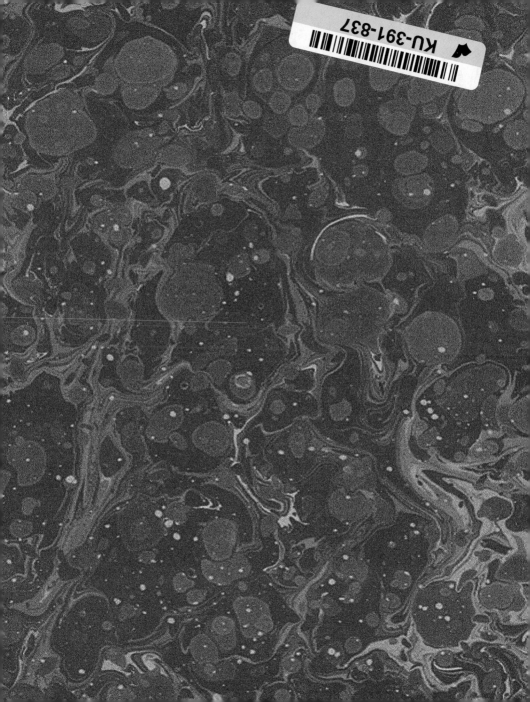

SCIENCIA

Simultaneously published in the UK
by Wooden Books Ltd, Glastonbury
and in the US by Bloomsbury USA, New York.

A CIP catalogue record for this book is available from the British Library.

ISBN: 978-1-907155-12-3

Visit Wooden Books' excellent web site at www.woodenbooks.com

3 5 7 9 10 8 6 4

The paper in this book is made from
wood grown in well-managed forests.

Designed and typeset by Wooden Books Ltd,
Glastonbury, Somerset, UK
Printed and bound in China

SCIENCIA

mathematics, physics, chemistry,
biology and astronomy for all

"… that which is below is as that which is above,
and that which is above is as that which is below,
to perform the miracles of the One Thing."

The Emerald Tablet of Hermes Trismegistus

CONTENTS

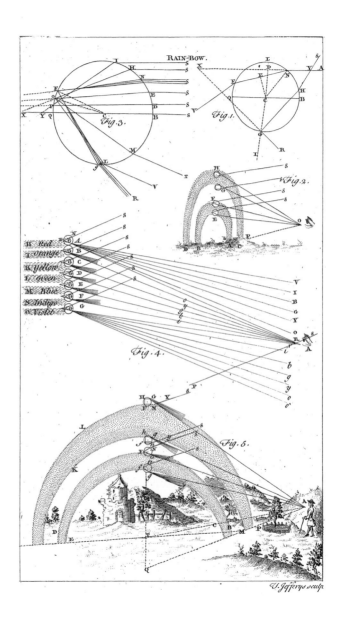

RAIN-BOW.

Fig. 3.

Fig. 1.

Fig. 2.

H *Red*
I *Orange*
Y *Yellow*
G *Green*
M *Blue*
S *Indigo*
V *Violet*

V
I
B
G
Y
O
R
A

Fig. 4.

Fig. 5.

T. Jefferys sculp.

EDITOR'S PREFACE

This volume contains six popular scientific titles from the Wooden Books series. With some sections completely reworked and over 40 new pages, it summarizes much of the maths, physics, chemistry, biology, and astronomy that any budding scientist or layperson should know.

First up is Prof. Burkard Polster's wonderful book on Mathematical Proofs *Q.E.D.*, to remind us that some things are just plain obvious. Second is Dr Matthew Watkins' dense collection of *Useful Mathematical and Physical Formulæ*, a good test for anyone. For thirds, we turn to chemistry in Matt Tweed's fizzy guide to the Periodic Table of *Essential Elements*. Fourth is Dr Gerard Cheshire's lively treatise on *Evolution*, the journey of life on Earth. Fifth, to deepen our understanding of biology, we examine one organism in more detail in Dr Moff Betts' beautiful guide to *The Human Body*. Finally, in the sixth book, we turn our eyes to the skies and ponder the story of the incredible universe we inhabit and are a part of, in Matt Tweed's tour through *The Compact Cosmos*.

Illustration credits for *Sciencia* include: Cecily Kate Borthwick, Allan Brown, Vivien Martineau, David Goodsell, Caroline Ede, Joe Mclaren, Dan Goodfellow, Will Spring, Simon Huson, NASA, Fermilab, and numerous engravers from centuries past. Other editors and designers have included Dr Peter Spring, Daud Sutton, Polly Napper, George Gibson, Mike O'Connor, and Dr Justin Avery.

Apologies to purists but a mix of both UK and US spelling and punctuation has been used throughout this jointly printed edition.

Thanks to all, and happy reading.

John Martineau

BOOK I

area triangle = area rectangle = 1/2 base · height

Q.E.D.

BEAUTY IN MATHEMATICAL PROOF

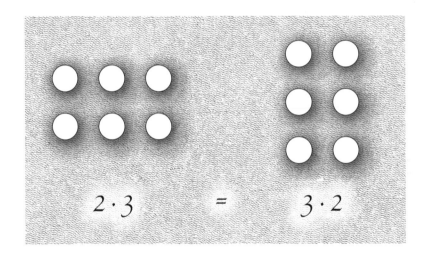

$$2 \cdot 3 \qquad = \qquad 3 \cdot 2$$

Burkard Polster

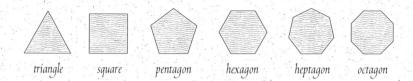

triangle   square   pentagon   hexagon   heptagon   octagon

A regular polygon is a convex figure with equal sides and angles. There are infinitely many regular polygons.

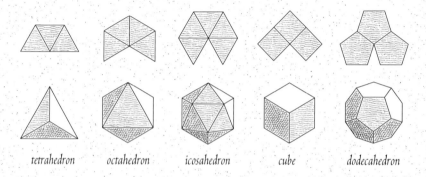

tetrahedron   octahedron   icosahedron   cube   dodecahedron

A regular polyhedron is a convex body with identical regular polygons as faces and the same number of faces meeting at every corner. Shown at the top are the different ways of joining three or more identical regular polygons to a corner with space left to fold up into three dimensions. These possibilities of building spatial corners then can be shown to extend in a unique way to the famous five regular polyhedra.

The same simple reasoning shows that there are three tilings of the plane with identical regular polygons.

Introduction

There are some mathematical objects whose beauty everyone is able to appreciate. The regular polygons and polyhedra are good examples—these figures are surpassed in perfection only by the circle and the sphere. Then there is Pythagoras' theorem, a cornerstone of the right-angled worlds we build for ourselves, and perhaps the conic sections which describe the orbits of celestial bodies.

Very few people appreciate more than some elementary aspects of mathematical beauty, much of it revealing itself only to mathematicians in the study and creation of intricately crafted proofs, barely within the reach of the most highly trained human minds.

As a mathematician, I declare that I have established the truth of a theorem by writing at the end of its proof the three letters Q.E.D., an abbreviation for the Latin phrase *quod erat demonstrandum*, which translates as "what had to be proved". On the one hand, 'Q.E.D.' is a synonym for truth and beauty in mathematics; on the other hand, it represents the seemingly inaccessible side of this ancient science.

'Q.E.D.' can, however, also be found at the end of some simple, striking, and visually appealing proofs. This little book presents a journey through a collection of these wondrous gems, exploring the ideas behind mathematical proof on the way, written for all those who are interested in the beauty of mathematics hidden below the surface.

TREACHEROUS TRUTH
what proofs are all about

In mathematics, as in the physical sciences, we may run an experiment or check a few cases to come up with a conjecture for a theorem. However, in mathematics experiments cannot replace proof, no matter how natural and obvious the conjecture is that they support. For example, the maximum number of regions defined by 1, 2, 3, 4, 5, and 6 points on a circle (*below*) are 1, 2, 4, 8, 16, and ... 31, and not 32!

Or, take the famous Goldbach conjecture which claims that every even number greater than two is the sum of two prime numbers as, for example, $12 = 5 + 7$ or $30 = 23 + 7$. Although this conjecture has been checked for many millions of cases, unless a proof is found, we cannot be sure that the next case we check won't show that the conjecture is false.

Proofs should be as short, transparent, elegant, and insightful as possible. Our proof (*opposite top*) that the number 0.999..., with infinitely many 9s, is equal to 1 is of this kind and its main argument can be easily adapted to convert any decimal number with one of those slightly worrying infinitely repeating tails into a number we are more comfortable with. The proof that the indented chess board cannot be tiled with dominos (*lower opposite*) is another example. Of course, the argument here applies to many other mutilated chess boards.

THEOREM: $1 = 0.9999\ldots$

PROOF: SET $X = 0.999\ldots$ THEN

$$10X = 9.999\ldots$$
$$-X = 0.999\ldots$$

$$= 9X = 9.000\ldots$$

THUS $X = 1.000\ldots$

Q.E.D

 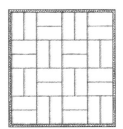 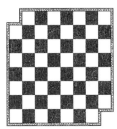

A chess board can be tiled with dominos each covering two squares. However, the indented board cannot.

Proof: In any tiling, a domino covers one white and one black square. Hence a tiling covers the same number of white and black squares. Since the indented board has two more white squares than black squares, it cannot be tiled. Q.E.D.

PYTHAGORAS' THEOREM
a proof by dissection

The Theorem of Pythagoras (*ca.* 569–475 BC) states that in a right-angled triangle the square on the hypotenuse, or long side, is equal to the sum of the squares on the other two sides (*opposite top*). Nowadays this is written algebraically as $a^2 + b^2 = c^2$.

PROOF: Arrange four identical right-angled triangles with sides a, b and c in a large square of side $a+b$, leaving two square spaces with sides a and b, respectively (*center opposite left*). The four triangles can also be arranged in this large square to leave a central square space with side c (*center opposite right*). In both cases the contained squares equal the large square minus four times the triangle. Therefore the sum of the two smaller squares, $a^2 + b^2$, equals the larger square, c^2. Q.E.D.

Conversely, and this requires an extra proof, IF a triangle's sides are related as above, THEN it is right-angled. Integers which satisfy the equality $a^2 + b^2 = c^2$ are known as Pythagorean triples. An ancient construction of a right angle from a loop of string with $3+4+5=12$ equally spaced knots is based on the triple $3:4:5$ (*below left*). A Babylonian clay tablet, Plimpton 322, lists integer pairs corresponding to Pythagorean triples (*below right*), which suggests that our general theorem may well have been known long before Pythagoras.

$$65^2 + 72^2 = 97^2$$
$$119^2 + 120^2 = 169^2$$
$$319^2 + 360^2 = 481^2$$
$$2291^2 + 2700^2 = 3541^2$$

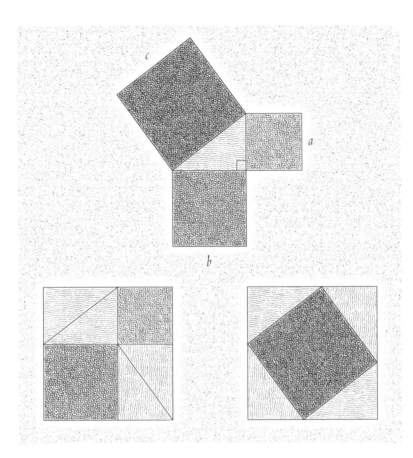

If, instead of squares, we fit any three similar figures to the sides of a right-angled triangle, we can also prove that the areas of the smaller two add up to that of the largest.

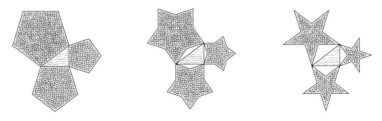

PLANE AND SIMPLE
your basic theorem toolbox

The *Elements* of Euclid (*ca.* 325–265 BC) long ago set the standard for mathematical rigour; and, being a popular textbook ever since, much of its contents have been absorbed into our common cultural heritage.

Over the course of thirteen books, Euclid built up a complex network of theorems of ever increasing depth, connected by logical arguments and rooted in some intuitive facts, called *axioms* or *postulates*. To be prepared for the rest of this book, start with the four simple results on the right and, following the arrows, deduce the theorems on the left.

You also need to be able to recognize in a flash the two main levels of sameness of triangles. Two triangles are *similar* if they have equal angles. Since two angles in a triangle determine the third, you know that two triangles are similar if you can show that they share two angles. Two triangles are *congruent* if they have equal sides. This is the case whenever one of the five configurations of three sides and angles drawn solid below are present in both triangles. For example, in the diagram on the right (*below*), the two gray triangles share one such configuration consisting of the two sides *r* and *m* and one right angle and are therefore congruent. Hence the two tangent segments *s* and *t* to the circle from the point outside have the same length.

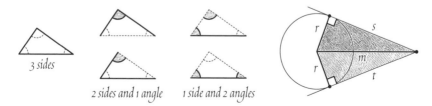

3 sides

2 sides and 1 angle   1 side and 2 angles

The Sum of the Angles in a Triangle

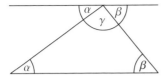

The sum of the angles is $\alpha + \beta + \gamma = 180^o$.

If the lines k and l are parallel

then $\alpha = \beta$

If

then $\alpha + \beta + \gamma = 180^o$

Thales' Theorem

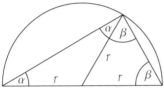

The upper angle $\alpha + \beta$ is 90°.

If $a = b$

then $\alpha = \beta$, and vice versa

Squaring a Rectangle

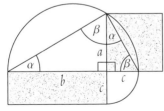

area square $= a^2 = b \cdot c =$ area rectangle
(by similarity of triangles $a/c = b/a$).

In similar triangles

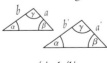

$a/a' = b/b'$

13

FROM PIE TO PI
mysteries of the circle

Eratosthenes (276–194 BC) is famous for his pizza pie method for calculating the circumference of the Earth based on the distance from Alexandria to Syene and the shadow angle at Alexandria at a time when the sun shone down a deep well in Syene, throwing no shadow there. Using the formula *circle diameter* × π = *circle circumference*, he also calculated the Earth's diameter. Fortunately, his penpal Archimedes (287–212 BC) was able to prove a good estimate for the elusive value of π.

Since π is the circumference of a circle of diameter 1, it will be greater than the circumference of any inscribed and less than that of any circumscribed regular polygon (*opposite top*). The more sides such a polygon has, the closer its circumference will be to that of the circle. Luckily, it is easy to calculate from the circumference of one such polygon the circumference of a polygon of the same kind with double the number of sides (*center opposite*). Starting with regular hexagons, Archimedes successively calculated the circumferences of regular 12-, 24-, 48- and 96-gons to capture π between $3\frac{10}{71}$ and $3\frac{10}{70}$. The last of these values equals $\frac{22}{7}$ and is used even today in many school books instead of the true value of π. Using squares instead of hexagons, a formula for approximating π emerges (*lower opposite*).

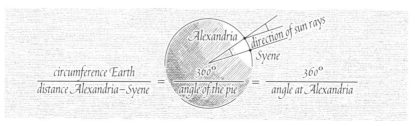

$$\frac{circumference\ Earth}{distance\ Alexandria-Syene} = \frac{360°}{angle\ of\ the\ pie} = \frac{360°}{angle\ at\ Alexandria}$$

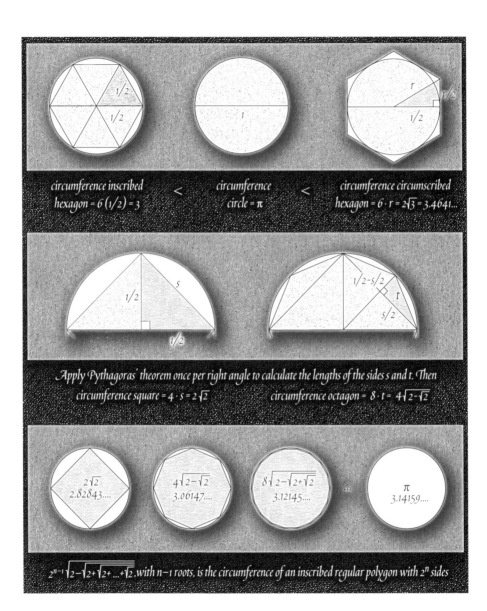

circumference inscribed
hexagon = 6 (1/2) = 3

<

circumference
circle = π

<

circumference circumscribed
hexagon = 6 · r = 2√3 = 3.4641...

Apply Pythagoras' theorem once per right angle to calculate the lengths of the sides s and t. Then

circumference square = 4 · s = 2√2

circumference octagon = 8 · t = 4√(2-√2)

2√2
2.82843....

4√(2-√2)
3.00147....

8√(2-√2+√2)
3.12145....

...

π
3.14159....

$2^{n-1}\sqrt{2-\sqrt{2+\sqrt{2+...+\sqrt{2}}}}$,with n−1 roots, is the circumference of an inscribed regular polygon with 2^n sides

15

CAVALIERI'S PRINCIPLE
a proof by approximation in slices

There are two versions of the celebrated principle named after Bonaventura Cavalieri (1598–1647). For plane figures it says that if every horizontal line intersects two plane figures in cuts of equal length, then the two figures will have equal area. Similarly, if every horizontal plane intersects two solids in cuts of equal area, then the two solids will have equal volume.

An outline of the proof by approximation in slices, which is the same for both principles, is given on the opposite page. Cavalieri's principle is a good example of "divide (into manageable pieces) and conquer" in mathematics. For example, in the plane version, we reduce the difficult problem of calculating areas to the easier problem of measuring the lengths of line segments.

Below are some important area and volume formulæ easily derived using Cavalieri's principle.

area parallelogram = area rectangle with the same base and height = base · height
area triangle = 1/2 area parallelogram = 1/2 base · height

volume prism or cylinder = volume box with the same base area and height = base area · height

every horizontal intersects the two figures in cuts of equal length

therefore, these two rectangles are congruent

therefore, the two piles of rectangles have the same area

finer cuts yield piles whose area is close to that of the originals

infinitely many cuts

CAVALIER CONE CARVING
serious dissection in action

Cones come in all sorts of shapes and sizes: Piles of sand, limpet shells, pyramids, church spires, crystal tips and unicorn horns are all examples of cones. Every cone has a vertex and a base which can be any plane figure.

Imagining the vertex as a beacon, a point is in the cone if its shadow is in the base. Let us prove that the formula for the volume of a cone is $\frac{1}{3} \times$ *base area* $\times$ *height*, implying the volume formulæ below.

A little shadow play (*opposite top*) illustrates that all cones of the same height and base area are cut by any horizontal plane in slices of equal area. Now, Cavalieri's principle (*see previous page*) tells us that ALL these cones have the same volume. It is therefore enough to calculate the volume of ONE of these cones such as that of the right-angled pyramid (*center opposite*). This and the other two pyramids combine into the triangular prism. Since all three pyramids have equal volume, this volume is one third of the volume of the prism. Q.E.D.

To cut a cube into six triangular pyramids of equal volume, first slice it with a diagonal plane into two triangular prisms and then these (as above) into three pyramids each. Or cut it into three identical square pyramids (*lower opposite*) and then each of these into a pyramid P_3 and its mirror image. These six pieces made from paper make a nice puzzle.

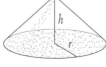

volume sand pile $= \frac{1}{3} \pi r^2 h$

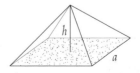

volume pyramid $= \frac{1}{3} a^2 h$

a punctual beacon projects
figures of equal area in a plane to
figures of equal area in any parallel plane
therefore the two cones have the same volume

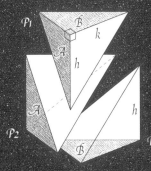 =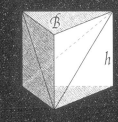

pyramids P_2 and P_1 share the same base A and height k
P_1 and P_3 also share base B and height h
volume of P_1 = volume of P_2 = volume P_3 = 1/3 volume prism = 1/3 base area · height = 1/3 $B · h$

 = 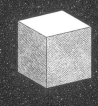 =

dissection of a cube into: (1) three identical square pyramids, (2) six triangular pyramids P_3

A FRUSTRATING FRUSTUM
horses and moat walls

Many ancient manuscripts contain algorithms for calculating the areas or volumes of geometric figures, but not all formulæ used by the ancients were correct. According to a Babylonian source, the volume of a frustum, or truncated square pyramid, was $(\frac{1}{2}(a+b))^2 h$, whereas the Egyptian Rhind papyrus (*ca.* 1800 BC), indicates the descendants of the pyramid builders were using the correct version $\frac{1}{3}(a^2 + ab + b^2)h$.

One of the oldest surviving Chinese mathematical treatises *Jiuzhang Suanshu* 九章算術 (*Arithmetic in Nine Chapters, ca.* 50 BC) also mentions this version of the formula, and Liu Hui 劉徽 (*ca.* 263 AD), in his commentary, gives a beautiful proof for it. He dissects the frustum or *fangting* 方亭 (square pavilion) into nine pieces: four identical pyramids or *yangma* 陽馬 (male horses), four prisms or *qiandu* 塹堵 (moat walls), and a rectangular box, all of which combine into a box and a pyramid. Adding the volumes of these together produces the volume of the frustum (*opposite top*).

This proof assumes that we already know the formula for the volume of a square pyramid (*see previous page*), but we can nevertheless use our frustum dissection to rediscover this in an elegant snake-bites-its-own-tail way (*lower opposite*).

Other solids' volumes from Liu Hui's commentary are shown below.

芻甍 *chumeng*
fodder loft

芻童 *chutong*
fodder boy

鱉腦 *bienao*
turtle's shoulder-joint

羨除 *xianchu*
drain

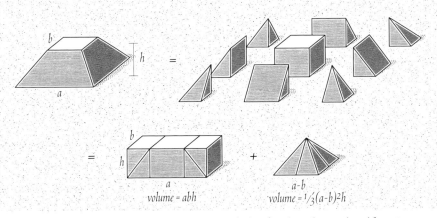

volume = abh   volume = $\frac{1}{3}(a-b)^2h$

Liu Hui dissects the frustum into nine pieces: a rectangular box, four identical pyramids, and four prisms, all of which combine into a box and a pyramid, with respective volumes abh and $\frac{1}{3}(a-b)^2h$. Adding these together produces the volume of the frustum, $\frac{1}{3}(a^2+ab+b^2)h$.

Doubling the linear size of any plane or solid shape results in a fourfold increase in area, or an eightfold increase in volume. Using this we can slice a pyramid halfway as shown below.

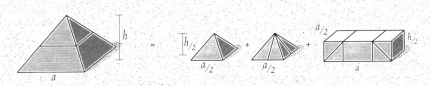

Volume of pyramid (v) = two $\frac{1}{8}$ volume pyramids + $\frac{a}{2} \cdot \frac{h}{2} \cdot a$
So, $\frac{3}{4}v = \frac{1}{4}a^2h$, which implies the volume of the pyramid, $v = \frac{1}{3}a^2h$.

Archimedes' Theorem

mysteries of the sphere

Archimedes proved that the volume of a sphere is two thirds that of the smallest cylinder containing it, and that its surface area is the same as that of the hollow cylinder. So taken was the philosopher with these relationships that he had a sphere and its surrounding cylinder inscribed on his tombstone.

Opposite we use Cavalieri's principle (*see page 16*) to derive the formula $\frac{4}{3}\pi r^3$ for the volume of a sphere of radius r, and thereby confirm Archimedes' first discovery.

For some real magic, project every point of the sphere, except the poles, onto another point on the cylinder, as shown below. Then it can be proved that any patch on the sphere gets projected onto a patch of equal area on the cylinder. If we then take the patch to be the whole sphere, it follows that its image is the cylinder, implying Archimedes' second discovery.

If we replace the sphere with a globe, project it onto the cylinder, and then slice the cylinder open, we find ourselves with a highly useful equal-area map of the Earth.

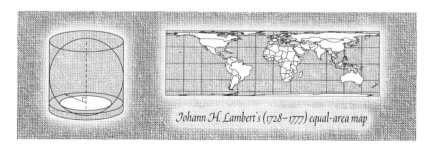

Johann H. Lambert's (1728–1777) equal-area map

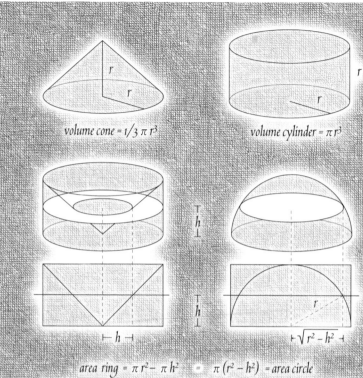

volume cone = $1/3\,\pi\,r^3$

volume cylinder = $\pi\,r^3$

area ring = $\pi\,r^2 - \pi\,h^2$ = $\pi\,(r^2 - h^2)$ = area circle

Since the area of the ring is the same as that of the circle, Cavalieri's principle yields that the hemisphere and the cylinder minus the cone have the same volume.

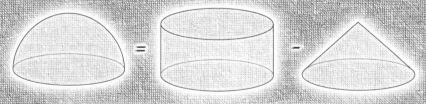

volume hemisphere = volume cylinder − volume cone = $\pi\,r^3 - 1/3\,\pi\,r^3 = 2/3\,\pi\,r^3$

volume sphere = 2 volume hemisphere = $4/3\,\pi\,r^3$.

INSIDE OUT
two proofs in wedges

Archimedes demonstrated how to mathematically kill two birds with one stone by using an ingenious idea to relate the insides and outsides of circles and spheres. Here is a sketch of his argument.

First he dissected the circle of radius *r* into a number of equal wedges (*opposite top*) and arranged them into a roughly rectangular slab. He then observed that this can be done with an ever increasing number of wedges, and that as the number of wedges grows, the slab becomes indistinguishable from a rectangle whose short side has length *r* and whose long side is half the circumference of the circle. Therefore the area of the rectangle coincides with that of the circle giving the formula:

$$area\ of\ circle = \tfrac{1}{2} \times circumference\ of\ circle \times r$$

We arrive at the same result by calculating the area of the sawtooth figure—just note that one of the 'triangles' has area $\tfrac{1}{2} \times base\ length \times r$, and that the bases of the triangles sum to the circumference.

To derive a similar formula for a sphere of radius *r*, Archimedes dissected it into triangular cones whose common vertex is the center of the sphere and whose bases are contained in the surface of the sphere (*lower opposite*). These cones can play the role of the triangles in the sawtooth figure, and since (*from page 18*) the volume of one of these ones is $\tfrac{1}{3} \times base\ area \times r$ we obtain the formula:

$$volume\ of\ sphere = \tfrac{1}{3} \times surface\ area\ of\ sphere \times r$$

As the grand finale, we plug the formulæ for the circumference of a circle of radius *r* and that of the volume of a sphere of radius *r* (*see previous page*) into our new relationships to conclude that the area of the circle is πr^2 and that the surface area of the sphere is $4\pi r^2$.

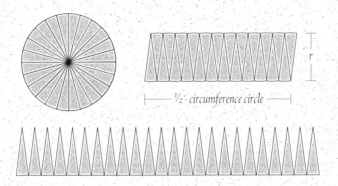

area of circle = area of wedges = ½ · circumference circle · r

volume of sphere = volume of cones = ⅓ · surface area sphere · r

MATHEMATICAL DOMINOS
proofs by induction

Setting up a number of dominos in a row, one domino for each natural number, let's make sure that IF domino n topples, THEN so does domino $n+1$. If we now topple the first domino, we can be sure that every domino will eventually topple over.

Proof by induction is the mathematical counterpart of this insight. Now, instead of the dominos, we have an infinite number of statements, one for each natural number. Here, we can be sure that all statements are true IF we can prove that the first statement is true AND the truth of statement n implies the truth of statement $n+1$.

The first three rows of diagrams (*opposite*) show how the first three statements corresponding to the following theorem imply each other:

THEOREM: Every 2^n by 2^n board that has been dented in one of its unit squares can be tiled with L-shapes made up of three unit squares.

PROOF BY INDUCTION: Since a dented 2 by 2 board is an L-shape (*opposite top*) the theorem is true for $n=1$. Assuming that statement n is true and considering an arbitrary dented 2^{n+1} by 2^{n+1} board, we quarter it and remove three middle squares to create four dented 2^n by 2^n boards (*center opposite*). By assumption these four boards can be tiled, and the four tilings extend to one of the 2^{n+1} by 2^{n+1} board. Q.E.D.

Some of the other tumbling patterns used in the art of domino toppling also translate into methods of proof. In the triangular pattern (*lower opposite*), for example, the front piece topples all other pieces. The corresponding method of proof can be used to show that Pascal's triangle, named after Blaise Pascal (1623–1662), is made up of binomial coefficients (*see too page 48*).

Every dented 2x2 board can be tiled with 'L' shapes

To show that an arbitrary dented 4x4 board can be tiled, quarter it, remove three middle squares to create four dented 2x2 boards and a tiling emerges. This suggests how to tackle the next case.

To show that an arbitrary dented 8x8 board can be tiled, quarter it and remove three middle squares to create four dented 4x4 boards. Extend tilings of these four to one of the 8x8 board.

$$1$$
$$1 \quad 1$$
$$1 \quad 2 \quad 1$$
$$1 \quad 3 \quad 3 \quad 1$$
$$1 \quad 4 \quad 6 \quad 4 \quad 1$$

$$\binom{0}{0}$$
$$\binom{1}{0} \quad \binom{1}{1}$$
$$\binom{2}{0} \quad \binom{2}{1} \quad \binom{2}{2}$$
$$\binom{3}{0} \quad \binom{3}{1} \quad \binom{3}{2} \quad \binom{3}{3}$$
$$\binom{4}{0} \quad \binom{4}{1} \quad \binom{4}{2} \quad \binom{4}{3} \quad \binom{4}{4}$$

One proof that the number triangles are equal corresponds to tumbling the pattern on the left. Just note that in both number triangles any entry is the sum of the entries above and $1 = \binom{0}{0}$.

THE INFINITE STAIRCASE
a proof by regrouping

A classical paradox involves a number of identical bricks that are stacked up on top of a desk, as in the diagram (*opposite*). It is easy to prove that by adding more and more bricks as indicated, we can make the resulting staircase protrude as much as we want without it toppling over.

A staircase of n bricks, each of length 2, protrudes a distance of

$$1 + \frac{1}{2} + \frac{1}{3} + \frac{1}{4} + \cdots + \frac{1}{n}$$

So, what we want to demonstrate is that the above sum approaches infinity as n does.

PROOF: We first group the infinite sum as follows:

$$1 + \frac{1}{2} + \left(\frac{1}{3} + \frac{1}{4}\right) + \left(\frac{1}{5} + \frac{1}{6} + \frac{1}{7} + \frac{1}{8}\right) + \left(\frac{1}{9} + \frac{1}{10} + \frac{1}{11} + \frac{1}{12} + \frac{1}{13} + \cdots\right)$$

We replace every term by a number less than or equal to it to produce a new sum which is less than or equal to the one we started with, and we notice that our substitutes add up to infinity:

$$1 + \frac{1}{2} + \left(\underbrace{\frac{1}{4} + \frac{1}{4}}\right) + \left(\underbrace{\frac{1}{8} + \frac{1}{8} + \frac{1}{8} + \frac{1}{8}}\right) + \left(\underbrace{\frac{1}{16} + \frac{1}{16} + \frac{1}{16} + \frac{1}{16} + \frac{1}{16}}\right) + \cdots$$

$$1 + \frac{1}{2} + \quad \frac{1}{2} \quad + \qquad \frac{1}{2} \qquad + \qquad \frac{1}{2} \qquad + \cdots$$

This means that the sum we are chasing is also infinite. Q.E.D.

Note that the staircase also gets infinitely tall as it grows infinitely broad and that actually building it gets very tricky very fast, involving closer and closer spacings of the bricks.

To construct a staircase of maximal overhang that does not topple over, we build from top to bottom such that at any step the centre of gravity of what we have built so far comes to rest on top of the front edge of the next brick.

Then, simple physics tells us that the indent x marking the position of the centre of gravity of the first n bricks as above satisfies the equation, $x \cdot (\text{weight } n\text{-}1 \text{ bricks}) = (1\text{-}x) \cdot (\text{weight one brick})$, or $x \cdot (n\text{-}1) = 1\text{-}x$. Solving for x gives $x = 1/n$.

CIRCLING THE CYCLOID
a proof by dissection

Start with a regular polygon under a line, mark one of its top corners, and start rolling it along the line. Every time the polygon comes to rest on the line, indicate the position of the marked corner by a dot. Stop when the coloured corner again touches the line and connect the dots by straight lines (*below left*). Dissecting the polygon, it quickly becomes clear that the area enclosed by the resulting curve is exactly three times the area of the polygon (*opposite top*).

Using a circle instead of a polygon, the resulting curve is a cycloid (*below right*), used (with its relatives) by the ancient Greeks to describe the orbits of the planets. Since a circle can be approximated by regular polygons, the area enclosed by the cycloid is also three times the area of the circle.

The cycloid has many other important properties. For example, demonstrating the formidable power of Newton's and Leibniz's new infinitesimal calculus, Johann Bernoulli proved in 1696 that the cycloid is the solution to the difficult classical 'problem of quickest descent'. This means that if a particle slides along the cycloid from one of its ends to a second point, driven by gravity alone, it does so in less time than along any other curve connecting the two points.

Puzzle out the two one-glance proofs (*lower opposite*)!

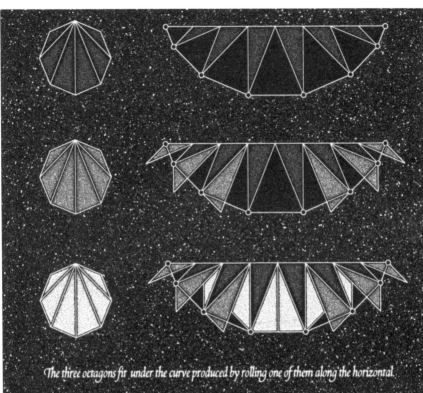

The three octagons fit under the curve produced by rolling one of them along the horizontal.

little square = 1/5 big square

area of regular 12-gon inscribed in circle of radius 1 is 3

SLICING CONES
Dandelin's sphere trick

What kind of curve do you get when you slice a circular cone with a plane into an upper and a lower part? It may seem counterintuitive, but this shape will always be an ellipse, that is, the kind of curve you get when you pin the two ends of a piece of thread to a desk, pull the thread taut with a pen and draw a closed curve (*below*). In other words, an ellipse is the set of all those points in the plane the sum of whose distances from two fixed points (the focal points) is a constant.

To prove the slicing-cone theorem, Germinal Dandelin (1794–1847) inscribed two spheres into the cone that touch the slicing plane at one point each (*opposite top*). He then observed that the cut is indeed an ellipse with these two points as focal points and associated constant the distance between the circles in which the two spheres touch the cone.

Similar tricks show that a plane cuts the cone in an ellipse, a parabola, a hyperbola (*lower opposite*) or, if it contains the vertex, a point, a line, or a pair of lines. Newton proved that two celestial bodies will always orbit each other on one of these conic sections, e.g. every planet orbits the Sun on an ellipse with the Sun at one of the focal points.

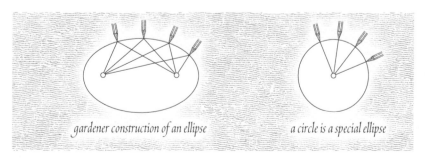

gardener construction of an ellipse   *a circle is a special ellipse*

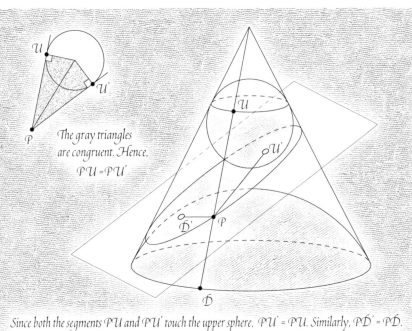

The gray triangles
are congruent. Hence,
PU = PU'

Since both the segments PU and PU' touch the upper sphere, PU' = PU. Similarly, PD' = PD. Therefore, PU'+PD'=PU+PD=UD, the distance between the touching circles of the spheres.

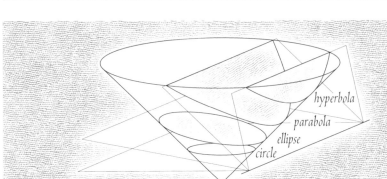

hyperbola
parabola
ellipse
circle

Tilting a plane up around the line, it first cuts the cone in a circle in the horizontal position, then ellipses, then a parabola in exactly one position, and from then on hyperbolae.

FOLDING CONICS
burning mirrors and whispering walls

Mark a dot on a circular piece of paper, fold a perimeter point onto the dot to produce a crease line, and repeat for various points on the perimeter. An ellipse will begin to appear (*opposite top*).

The proof that this works hinges on the defining property of ellipses in terms of two focal points (*see previous page*) and the property that motivates the term 'focal point': If we make the inside of an ellipse into a mirror, every light ray originating at one of its two focal points will pass through the other, having been reflected off the ellipse (*below left*). This is the principle behind burning mirrors and whispering walls. Put a candle at one of the focal points and its heat will focus at the other, whisper in one of the focal points of a large elliptical wall and your friend in the faraway other focal point will be able to hear you clearly. In general, light rays that miss the focal points envelope ellipses or hyperbolæ sharing the focal points with the initial ellipse (*below, middle and right*).

If you replace the circular piece of paper with a rectangular one and fold from one of its sides only (*lower opposite*), you get part of a parabola. From this, we can reconstruct the definition of a parabola in terms of a focal point and a line, and see how Archimedes came up with the idea of a parabolic array of mirrors focusing the sun to burn warships.

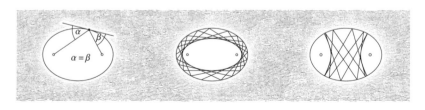

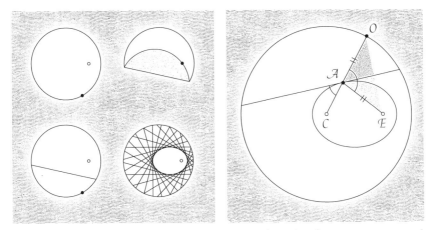

Since AO folds onto AE, AE + AC = AO + AC = CO = radius circle. Hence, as we move O around the circle, the point A describes an ellipse with focal points C and E and associated constant this radius. The three marked angles at A are equal. Therefore the creases are tangents of the ellipse and as such envelope it.

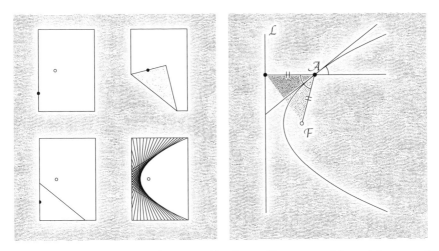

Folding of a parabola using its focal point F and base line L: (1) a point A of the parabola is at equal distance from F and L; (2) every horizontal light ray passes through F after being reflected off the parabola.

Knotting Polygons
a proof by paper folding

It is very easy to construct equilateral triangles, squares, and regular hexagons in a large number of different ways. Regular pentagons are more tricky, but here is the simplest way to construct one:

Tie a knot into a piece of tape and pull on the ends until the knot is completely flat. Cut off the excess tape on both sides and you are left with a regular pentagon! Why does this work?

Consider two regular pentagons sharing one side together with a piece of tape running through both of them (*below left*). If we fold the left pentagon onto the right along the common side, the paper strip will neatly align itself along one of the sides of the right pentagon. Therefore, if we keep folding the tape around this pentagon, we will successively define all its sides and diagonals. Unwrapping the now creased tape and discarding the pentagons, we can finally tie the tape into a knot and flatten it such that no new creases appear.

The regular polygons with more than five sides can also be knotted using one or two paper strips, but the practical execution of these constructions can get very awkward.

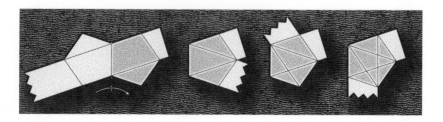

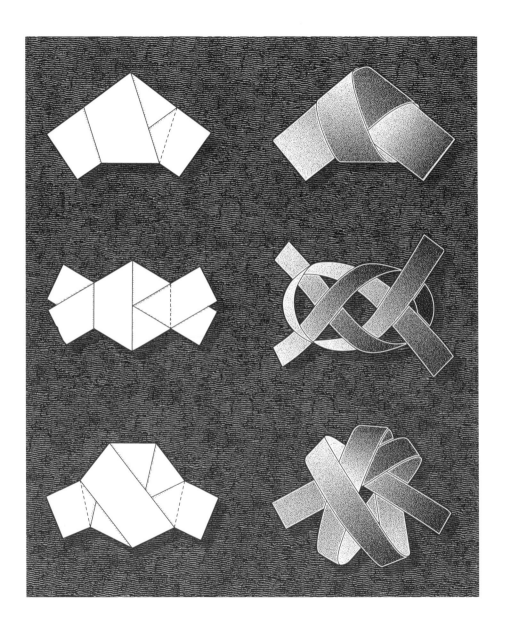

CUTTING SQUARES

cosmology and manifest number

Beautiful theorems are often arrived at by coming up with new ways of interpreting old patterns. For a whirlwind tour of some classic examples dating back to the Pythagoreans, let's consider various ways of dissecting a square array of n times n, or n^2, pebbles.

The first way gives the elementary equality $n + n + \ldots + n$ (n times) $= n^2$. The second way translates into the surprising fact that the sum of the first n odd numbers is equal to n^2. For another proof of this theorem, just note that the numbers of triangles in the columns of tiling n below are the first n odd numbers and that after separating the black and the gray triangles (*lower opposite*), we get a square of n^2 triangles.

Closely related is the third way of dissecting which corresponds to the equality $(n-1)^2 + (2n-1) = n^2$. Choosing the odd number $2n-1$ to be a square, we get a Pythagorean triple (*see page 10*). For example, choosing $2n-1 = 3^2$ gives $n = 5$, and therefore $4^2 + 3^2 = 5^2$.

One last way of dissecting the square array shows that n^2 is equal to the sum of the first n natural numbers plus the sum of the first $n-1$ natural numbers. Can you see how to derive from this a formula for the sum of the first n natural numbers?

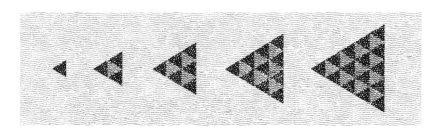

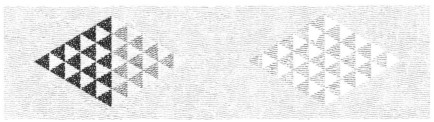

POWER SUMS
proofs by double counting

The marvellous Pythagorean one-glance dissection proof below shows that the sum of the first n natural numbers is half the number of pebbles in the rectangle, that is, $n(n + 1)/2$. Carl Friedrich Gauss (1777–1855), one of the giants of mathematics, rediscovered this formula at the age of ten. Asked by his teacher to sum up the first 100 natural numbers, he made short work of the tedious task by observing that,

$$1 + 100 = 2 + 99 = \ldots = 50 + 51 = 101,$$

and that therefore the required sum was $50 \times 101 = 5050$. This reasoning corresponds to looking carefully at the rectangle shown below, row by row $(1 + 4 = 2 + 3 = 5$ and the sum is $2 \times 5 = 10)$.

The first diagram on the right elegantly shows that three times the sum of the first n squares equals the number of pebbles in the rectangle, that is, $(2n + 1)(1 + 2 + \ldots + n)$.

The second diagram on the right demonstrates that the sum of the first n cubes is equal to the sum of the first n natural numbers squared.

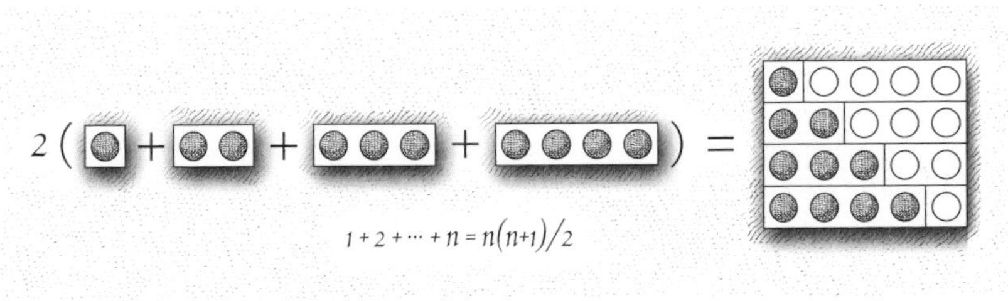

$$1 + 2 + \cdots + n = n(n+1)/2$$

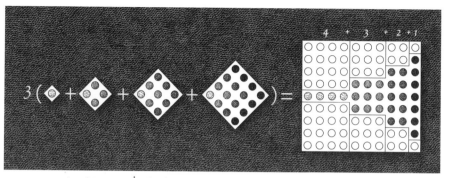

In general, $3\left(1^2 + 2^2 + \cdots + n^2\right) = \left(1 + 2 + \cdots + n\right)\left(2n + 1\right)$. Plugging in $1 + 2 + \cdots + n = n(n+1)/2$ gives
$$1^2 + 2^2 + \cdots + n^2 = n(n+1)(2n+1)/6$$

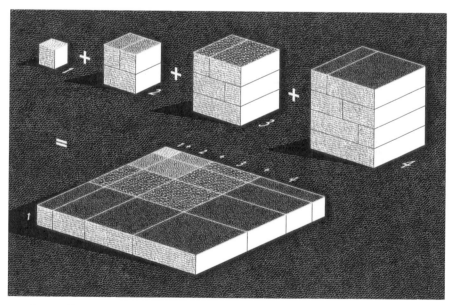

The sum of the volumes of the cubes, $1^3 + 2^3 + \cdots + n^3$, equals $1 \cdot \left(1 + 2 + \cdots + n\right)^2$, the volume of the square slab.
$$1^3 + 2^3 + \cdots + n^3 = \left(1 + 2 + \cdots + n\right)^2$$

NEVERENDING PRIMES
a proof by contradiction

Just as every object of the real world can be split in a unique way into indivisible atoms, every natural number can also be written in a unique way as the product of indivisibles called primes (the number 1 being an exception). The eight smallest primes are 2, 3, 5, 7, 11, 13, 17, and 19. The Sieve of Eratosthenes (*shown opposite*) is an elegant method for constructing all primes.

Euclid's *Elements* contains the following classic proof by contradiction that, unlike the real world, the world of numbers contains infinitely many primes.

PROOF: There are either finitely or infinitely many primes. Assume that there are only finitely many and multiply all of them together to form a very large integer $n = 2 \times 3 \times 5 \times 7 \dots$. Now, since $n+1$ is greater than any of the factors of n it cannot be prime, so one of the factors of n also has to be a factor of $n+1$. But, if this were so, then $(n+1) - n = 1$ would also have the same factor. This is a contradiction, so we conclude that our assumption of finitely many primes must be false. Hence there are infinitely many prime numbers. Q.E.D.

A *twin* of primes are two primes with a difference of two such as $5:7$ and $11:13$. Eternal fame awaits whoever can prove (or disprove) that there are infinitely many twins.

The Sieve of Eratosthenes

Circle all multiples of 2 in the above list. The smallest integer greater than 2 that is not circled is 3. Circle all its multiples. Now, the smallest integer greater than 3 that is not circled is 5. Circle its multiples, and so on. Then the primes are exactly the numbers that never get circled.

THE NATURE OF NUMBERS
another proof by contradiction

On the number line (*below*), every point represents one of the real numbers that we use for measuring distances, areas, and volumes. By dividing the intervals between the integers into two parts, three parts, four parts, and so on, we single out the fractions, or rational numbers.

Even the tiniest patch of the number line contains infinitely many rational numbers and it may therefore seem reasonable to expect that every real number is rational. The Pythagoreans reputedly sacrificed a hecatomb, or one hundred oxen, to celebrate the discovery of a proof that √2, the length of the diagonal of a unit square, is irrational, so not a rational number.

Our proof (*opposite*) is an example of a proof by contradiction. We start by assuming that √2 is rational. This first implies the existence of an integer square (a square with integer diagonals and sides) and eventually a contradiction, that is, a statement that is not true. We conclude that our assumption is false. Therefore, √2 is irrational.

In general, it can also be shown that if a natural number is not a square, then its square root is an irrational number. This means that infinitely many of the radii of the root spiral (*lower opposite*) are irrational. Also, it turns out that, in a sense, there are many more irrational numbers than there are rational ones.

44

If $\sqrt{2}$ was a fraction b/a of positive integers, then the above square inflated by the factor a is the integer square below (left). By Pythagoras' theorem, $a^2 + a^2 = 2\,a^2 = b^2$. Hence $a^2 = (b/2)b$ is an integer.

 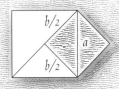

This is only possible if $b/2$ is an integer. Therefore, the small gray square on the right is also an integer square. Applying this construction to this second integer square gives a third, then a fourth, etc.

Every segment of the infinite zigzag on the right is a side of one of our integer squares and therefore has integer length. Impossible since the segments get infinitely small, whereas the smallest positive integer is 1.

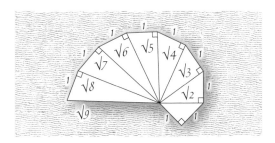

THE GOLDEN RATIO
nature's favourite number

What does a rectangle look like that is not too slim and not too wide, a rectangle that looks just right? For many artists and scientists this age-old beauty contest has a clear winner; the so-called golden rectangle (*below left*) whose ratio of long to short side equals the golden ratio ϕ (*phi*), of diagonal to side in a regular pentagon (*opposite top*).

The golden ratio is present in many of nature's designs such as leaf arrangements and spiral galaxies. For example, if we take away a square from a golden rectangle (*below*), we find we are left with another golden rectangle since $\phi = 1/(\phi-1)$ (*opposite top*). Repeating this process yields a spiral of squares that hugs many naturally occurring spirals.

Combining three golden rectangles at right angles (*center opposite*), their twelve corners become the corners of an icosahedron. To prove this, we only have to check that all the triangles in the middle picture are equilateral, or equivalently, that the essentially two different edges of these triangles all have equal length.

This paves the way to a beautiful construction of an icosahedron from an octahedron (*lower opposite*) where the twelve corners of the former divide the twelve edges of the latter in the golden ratio.

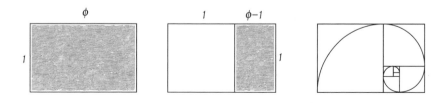

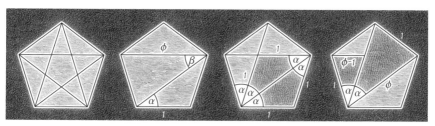

Angles α and β are equal. Therefore the first two gray triangles are congruent and the second two similar. Hence $\phi/1 = 1/(\phi - 1)$, or $\phi^2 = \phi + 1$. Solving this equation gives $\phi = (1+\sqrt{5})/2 = 1.61803...$

The highlighted edge AB has length $\sqrt{\phi^2+(\phi-1)^2+1^2} = \sqrt{2(\phi^2-\phi+1)}$ (apply Pythagoras' theorem twice) and since $\phi^2 = \phi + 1$, this is equal to 2, the length of the side of one of the three golden rectangles.

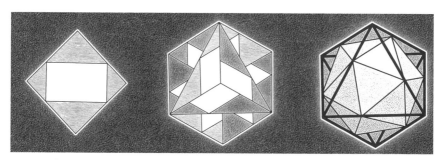

Put the three rectangles in squares as on the left. Then the edges of the squares form an octahedron.

THE NUMBERS OF NATURE
the geometry of growth

A spiral of squares growing around a unit square as on the top left consists of squares the lengths of whose sides are the Fibonacci numbers 1, 1, 2, 3, 5, 8, 13, 21, …, named after Leonardo Fibonacci (1170–1250). Every number in the sequence is the sum of the two preceding it, so that $2 = 1 + 1$, $3 = 1 + 2$, $5 = 2 + 3$, and so on.

Fibonacci numbers are connected in many wonderful ways. For example, the tiling of the rectangle on the top left demonstrates that $1^2 + 1^2 + 2^2 + 3^2 + 5^2 + 8^2 + 13^2 = 13 \times (13 + 8) = 13 \times 21$. In general, the sum of the squares of the first n Fibonacci numbers equals the product of the nth and $n + 1^{\text{st}}$ such numbers. Similarly, the tiling of the square on the right shows that $(1 \times 1) + (2 \times 1) + (3 \times 2) + (5 \times 3) + (8 \times 5) + (13 \times 8) + (21 \times 13) = 212$. This equality also generalizes easily.

The Fibonacci numbers often show up in the same phenomena as the golden ratio ϕ (*see previous page*) and it can be proved that the nth Fibonacci number is the closest natural number to $\phi^n / \sqrt{5}$. This implies that the rectangles we come across when building our spiral of squares become indistinguishable from golden rectangles.

Fibonacci numbers are hidden in many growth processes. For example, the numbers of clockwise and counterclockwise spirals apparent in sunflower heads (*center opposite*) are usually consecutive Fibonacci numbers. Pascal's triangle (*lower opposite*) also grows, here row by row, with neighboring entries in one row adding up to the number below them. Since the sums of the first two diagonals of this triangle are both 1, and the sums of any two consecutive diagonals add up to the sum of the next, our golden sequence will appear yet again.

The different stages of the infinite spiral of squares form rectangles (left) the lengths of whose sides are two consecutive Fibonacci numbers. The first n such rectangles tile a square (right) if n is an odd number.

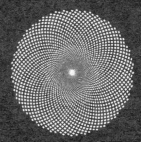

The numbers of clockwise and counterclockwise spirals visible in a sunflower head are (usually) consecutive Fibonacci numbers such as 34/21 on the left and 89/55 on the right.

```
              1
            1   1
          1   2   1
        1   3   3   1
      1   4   6   4   1
    1   5   10  10  5   1
  1   6   15  20  15  6   1
1   7   21  35  35  21  7   1
```

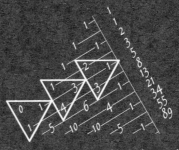

Every entry in a diagonal is the sum of one entry each from the previous two diagonals as on the right. This shows that the sums of two consecutive diagonals add up to the sum of the following diagonal.

EULER'S FORMULA
a proof by pruning

A cut diamond is a solid without indentations all of whose faces are flat polygons. Leonhard Euler (1707–1783) discovered the neat formula that relates the numbers of vertices, edges, and faces of such a solid:

$$V(ertices) + F(aces) - E(dges) = 2$$

For example, in the case of a cube we count 8 vertices, 6 faces, and 12 edges and, indeed, $V+F-E = 8+6-12 = 2$.

PROOF: Start by opening out the network of vertices and edges to get a plane picture of the solid (*opposite top*) in the form of a map with the same numbers of vertices, edges, and faces (the outside counts as one face). We notice that inserting a diagonal into a face yields a map with the same $V+F-E$ (*opposite, second row*) and so insert diagonals until a map consisting solely of triangles is produced. Finally, working around the outer border of the map and eliminating one triangle at a time (*opposite, lower two rows*) we are left with a map consisting of just one triangle (3 vertices, 2 faces, 3 edges). Since at every step the value of $V+F-E$ does not change, then $V+F-E = 3+2-3 = 2$. Q.E.D.

It is also not hard to show that Euler's formula holds true for any connected plane network of vertices and curve segments (*below*).

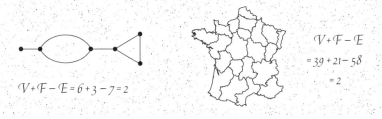

$V+F-E = 6+3-7 = 2$

$V+F-E$
$= 39 + 21 - 58$
$= 2$

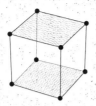 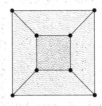

Opening out a cube gives a map with the same $V + F - E$.

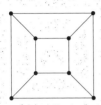 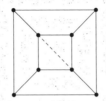 

Inserting a diagonal ups both the number of faces and edges by one leaving $V + F - E$ unchanged. Hence the $V + F - E$ of the triangulated map is the same as that of the original.

 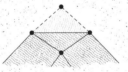

Pruning a triangle from the outside of a triangulated map leaves $V + F - E$ unchanged. For example, in the left diagram we lose one edge and one face: $V + F - E = V + (F - 1) - (E - 1)$.

Serious pruning shows that $V + F - E$ of the original solid equals that of a single triangle.

POSSIBLE IMPOSSIBILITIES
doubling, squaring, and trisecting

Socrates (469–399 BC) once used the first two diagrams below to show how to double a square, and the oracle of Delphi predicted that a plague could be stopped by doubling the cubical altar of Apollo.

In the nineteenth century it was proved that 'doubling a cube' as well as the other two notorious geometry problems of 'squaring a circle' and 'trisecting a general angle' are impossible if we require, like the ancient Greeks, that we use only a compass and unmarked ruler. If, however, we are allowed to use other tools, all three problems can be solved.

To square a circle, roll it half a revolution on a horizontal (*opposite top*) to construct the long rectangle which has the same area as the circle (*see page 25*). Now, using compass and ruler as indicated, construct the square which has exactly the same area as this rectangle (*see page 13*).

Archimedes discovered an ingenious method for trisecting an angle α between two intersecting lines (*center opposite*) using a compass and a ruler with two marks on it: Just draw a circle and align the ruler as in the diagram. Then the angle ε is exactly one third of the angle α.

Doubling a square amounts to constructing $\sqrt{2}$ from 1 (*below right*) and doubling a cube involves constructing $\sqrt[3]{2}$ from 1 (*lower opposite*). Just trisect and fold the paper unit square, as indicated, to construct $\sqrt[3]{2}$. Easy to describe but tricky to prove. Can you do it?

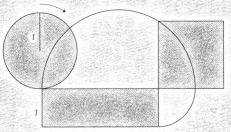

1/2 · circumference circle

area circle = area rectangle = area square

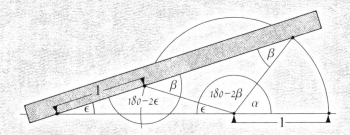

The first semicircle gives $\beta=180-(180-2\epsilon)=2\epsilon$ and the second $\alpha=180-(\epsilon+(180-2\beta))=3\epsilon$.

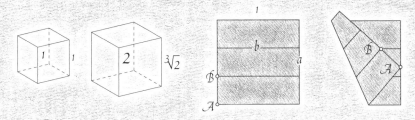

Folding $\mathcal{A}$ and $\mathcal{B}$ onto the segments a and b results in $\mathcal{A}$ dividing the segment a in the ratio $1:\sqrt[3]{2}$.

BOOK II

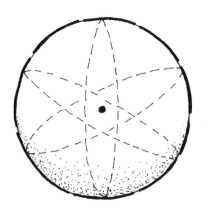

USEFUL
MATHEMATICAL AND PHYSICAL
FORMULÆ

Matthew Watkins

illustrated by Matt Tweed

INTRODUCTION

This little book is an attempt to present the basic formulæ of mathematics and physics in a friendly and usable form. Unfamiliar terms and symbols are defined in the index.

The use of number and symbol to model, predict, and manipulate reality is a powerful form of sorcery. Unfortunately, possession of these abilities doesn't necessarily bring with it wisdom or foresight. As a result, we see the proliferation of dangerous technologies and an increasing obsession with *quantity,* typified by the subservience of almost everything to the global economy. Readers are advised to use the contents of this book with due care and attention.

On the other hand, mathematical tools have made it possible to perceive unity in seemingly separate areas. Light and electricity, for example, once unrelated topics, are now both understood in terms of the theory of *electromagnetic fields.*

Brilliantly illustrating this 'double-edged sword', Einstein's well-known $E = mc^2$, perhaps the most famous formula of all, will be forever linked to both the creation of atomic weapons and the scientific (re)discovery of the unity of matter and energy.

May your senses of awe and delight never be extinguished!

TRIANGLES

and their various centers

A *right-angled* triangle obeys *Pythagoras' Law*: the square of the *hypotenuse* (the side opposite the right angle) is equal to the sum of the squares of the other two sides (*opposite top left*).

$$a^2 + b^2 = c^2 \quad \text{or equivalently} \quad c = \sqrt{a^2 + b^2}$$

The sum of internal angles in any triangle = 180°, or π *radians*.

The *perimeter* $\quad P = a + b + c$

The area $\quad S = \frac{1}{2}bh = \frac{1}{2}ab\sin C \quad$ (*see opposite top right*).

Sine Rule: $\quad \dfrac{a}{\sin A} = \dfrac{b}{\sin B} = \dfrac{c}{\sin C} = 2r$

(where r is the radius of the *circumcircle*).

Medians connect corners to the midpoints of the opposite sides. The three medians meet at a point called the *centroid*:

$$m^a = \frac{1}{2}\sqrt{2(b^2 + c^2) - a^2} \qquad m^b = \frac{1}{2}\sqrt{2(a^2 + c^2) - b^2}$$

$$m^c = \frac{1}{2}\sqrt{2(a^2 + b^2) - c^2}$$

Altitudes are line segments drawn from each corner to the opposite side (or its extension), meeting it at a right angle:

$$h_a = \frac{2S}{a} \qquad h_b = \frac{2S}{b} \qquad h_c = \frac{2S}{c}$$

The three altitudes meet at the *orthocenter*.

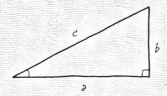

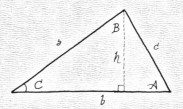

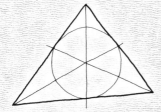

a triangle's angular bisectors meet at the center of its incircle

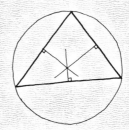

a triangle's perpendicular bisectors meet at the center of its circumcircle

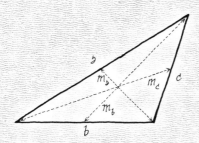

medians are drawn to the midpoints of each side and give the center of gravity of a uniform sheet

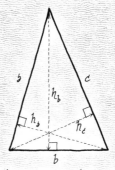

altitudes intersect at the orthocenter (not necessarily inside the triangle)

TWO-DIMENSIONAL FIGURES
areas and perimeters

Formulæ for the perimeters and areas of various two-dimensional forms are given below.

Circle: Radius = r, diameter $d = 2r$
Perimeter, or *circumference* = $2\pi r = \pi d$
Area = πr^2 where $\pi = 3.1415926\ldots$

Ellipse: Area = πab
a and b are the *minor* and *major semi-axes* respectively. The two illustrated points are its *foci*, such that $l + m$ is constant.

Rectangle: Area = ab
Perimeter = $2a + 2b$

Parallelogram: Area = $bh = ab \sin \alpha$
Perimeter = $2a + 2b$

Trapezium: Area = $\frac{1}{2}h(a + b)$
Perimeter = $a + b + h(\csc\alpha + \csc\beta)$

Regular n-gon: Area = $\frac{1}{4}nb^2 \cot(180°/n)$
Perimeter = nb
Sides and internal angles are all equal.

Quadrilateral (i): Area = $\frac{1}{2}ab \sin\alpha$

Quadrilateral (ii): Area = $\frac{1}{2}(h_1 + h_2)b + \frac{1}{2}ah_1 + \frac{1}{2}ch_2$

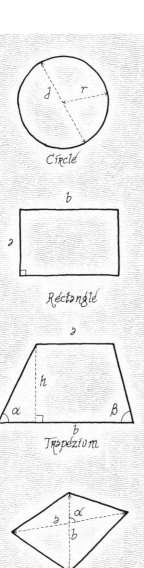

Circle

Ellipse

Rectangle

Parallelogram

Trapezium

Regular n-sided polygon

Quadrilateral (i)

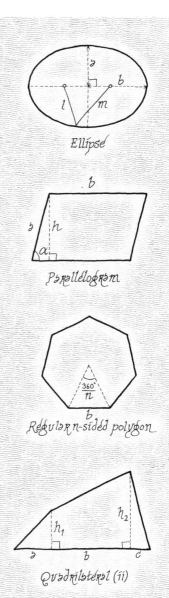

Quadrilateral (ii)

63

THREE-DIMENSIONAL FIGURES
volumes and surface areas

Formulæ for the volumes and surface areas (including bases) of eight three-dimensional solids are given below.

Sphere: Volume = $\frac{4}{3}\pi r^3$
Surface area = $4\pi r^2$

Box: Volume = abc
Surface area = $2(ab + ac + bc)$

Cylinder: Volume = $\pi r^2 h$
Surface area = $2\pi rh + 2\pi r^2$

Cone: Volume = $\frac{1}{3}\pi r^2 h$
Surface area = $\pi r\sqrt{r^2 + h^2} + \pi r^2$

Pyramid: Base area A
Volume = $\frac{1}{3}Ah$

Frustum: Volume = $\frac{1}{3}\pi h(a^2 + ab + b^2)$
Surface area = $\pi(a + b)c + \pi a^2 + \pi b^2$

Ellipsoid: Volume = $\frac{4}{3}\pi abc$

Torus: Volume = $\frac{1}{4}\pi^2(a + b)(b - a)^2$
Surface area = $\pi^2(b^2 - a^2)$

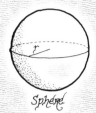

Sphere

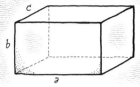

Box (Rectangular paralleliped)

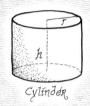

Cylinder

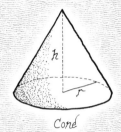

Cone

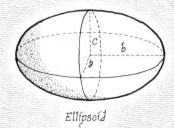

Pyramid with polygonal base, area A

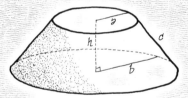

Frustum (truncated cone)

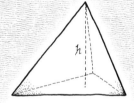

Ellipsoid

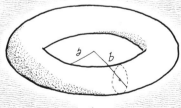

Torus

COORDINATE GEOMETRY
axes, lines, gradients and intersections

A pair of axes imposed on the plane at right angles allow a point to be defined by a pair of real numbers (*opposite*). The axes intersect at $(0,0)$, the *origin*. Horizontal and vertical positions are often referred to as x and y respectively.

The equation of a line is given by $y = mx + c$ where m is its gradient. This line cuts the y-axis at $(0, c)$ and the x-axis at $\left(-\frac{c}{m}, 0\right)$.

A vertical line has a constant x value, taking the form $x = k$.

The line passing through (x_0, y_0) with gradient n is given by the equation $y = nx + (y_0 - nx_0)$. A line perpendicular to another of gradient n will have gradient $-\frac{1}{n}$.

The equation of the line through (x_1, y_1) and (x_2, y_2) is,

$$y = \left(\frac{y_2 - y_1}{x_2 - x_1}\right)(x - x_2) + y_2 \quad \text{when } x_1 \neq x_2$$

The angle θ between two lines, gradients m and n, satisfies,

$$\tan \theta = \frac{m - n}{1 + mn}$$

A circle, radius r, center (a, b) is given by $(x - a)^2 + (y - b)^2 = r^2$.

In three dimensions, a z-axis is added and many equations take analogous forms. For instance, a sphere with radius r and centered at (a, b, c) is given by $(x - a)^2 + (y - b)^2 + (z - c)^2 = r^2$. The general equation for a three-dimensional plane is $ax + by + cz = d$.

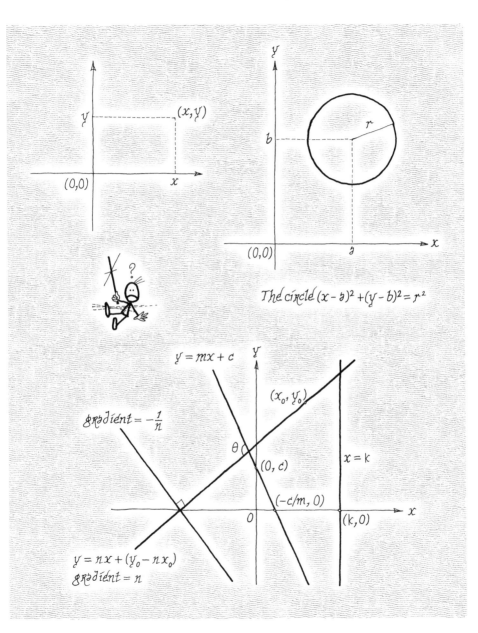

(x, y)

y

$(0,0)$ x

y

r

b

$(0,0)$ a x

The circle $(x - a)^2 + (y - b)^2 = r^2$

$y = mx + c$ y

(x_0, y_0)

gradient $= -\dfrac{1}{n}$

θ

$(0, c)$ $x = k$

$(-c/m, 0)$

0 $(k, 0)$ x

$y = nx + (y_0 - nx_0)$
gradient $= n$

TRIGONOMETRY
applied to right triangles

A right triangle with sides of length *a, b, c,* and angle θ is shown opposite left. The six *trigonometric functions*: sine, cosine, tangent, cosecant, secant, and cotangent are then defined as follows:

$$\sin\theta = \frac{b}{c} \qquad \cos\theta = \frac{a}{c} \qquad \tan\theta = \frac{b}{a}$$

$$\csc\theta = \frac{c}{b} \qquad \sec\theta = \frac{c}{a} \qquad \cot\theta = \frac{a}{b}$$

The sine and cosine are the height and base of a right triangle in a circle of radius 1 as shown below and opposite top right:

$$a = \cos\theta \quad \text{and} \quad b = \sin\theta$$

By Pythagoras' Theorem (*see pages 10 & 60*) we know $a^2 + b^2 = c^2$, hence for any angle θ in the circle we have the important identity,

$$\cos^2\theta + \sin^2\theta = 1$$

Sines, cosines and tangents have positive or negative values in different quadrants of the circle as shown below right.

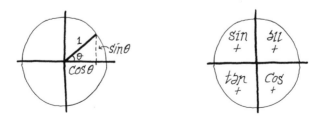

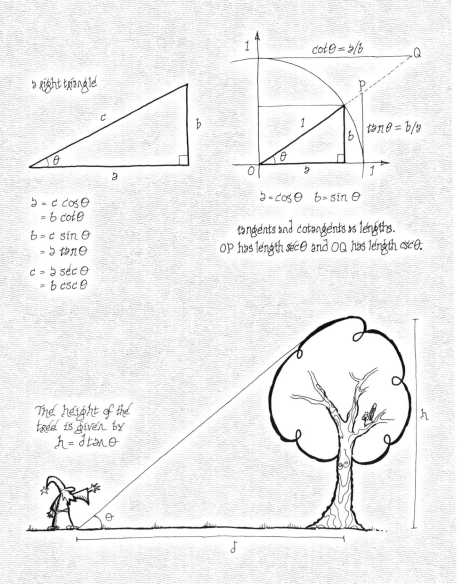

a right triangle

$$a = c \cos\theta$$
$$= b \cot\theta$$
$$b = c \sin\theta$$
$$= a \tan\theta$$
$$c = a \sec\theta$$
$$= b \csc\theta$$

$$\cot\theta = a/b$$

$$\tan\theta = b/a$$

$$a = \cos\theta \quad b = \sin\theta$$

tangents and cotangents as lengths.
OP has length $\sec\theta$ and OQ has length $\csc\theta$.

The height of the tree is given by
$$h = d \tan\theta$$

TRIGONOMETRIC IDENTITIES
relating the six functions

The definitions on the previous page lead to the following:

$$\tan\theta = \frac{\sin\theta}{\cos\theta}, \quad \cot\theta = \frac{\cos\theta}{\sin\theta}, \quad \sec\theta = \frac{1}{\cos\theta}, \quad \csc\theta = \frac{1}{\sin\theta}$$

Dividing $\cos^2\theta + \sin^2\theta = 1$ by $\cos^2\theta$ and $\sin^2\theta$ produces:

$$1 + \tan^2\theta = \sec^2\theta \qquad \text{and} \qquad 1 + \cot^2\theta = \csc^2\theta$$

The six functions applied to negative angles produce:

$$\sin(-\theta) = -\sin\theta \qquad \cos(-\theta) = \cos\theta \qquad \tan(-\theta) = -\tan\theta$$

$$\csc(-\theta) = -\csc\theta \qquad \sec(-\theta) = \sec\theta \qquad \cot(-\theta) = -\cot\theta$$

Angle sum formulæ apply when two angles are combined:

$$\sin(\alpha + \beta) = \sin\alpha\cos\beta + \cos\alpha\sin\beta$$

$$\cos(\alpha + \beta) = \cos\alpha\cos\beta - \sin\alpha\sin\beta$$

$$\tan(\alpha + \beta) = \frac{\tan\alpha + \tan\beta}{1 - \tan\alpha\tan\beta}$$

For doubled or tripled angles, use *multiple angle formulæ*:

$$\sin 2\alpha = 2\sin\alpha\cos\alpha \qquad\qquad \sin 3\alpha = 3\sin\alpha\cos^2\alpha - \sin^3\alpha$$

$$\cos 2\alpha = \cos^2\alpha - \sin^2\alpha \qquad\qquad \cos 3\alpha = \cos^3\alpha - 3\sin^2\alpha\cos\alpha$$

$$\tan 2\alpha = \frac{2\tan\alpha}{1 - \tan^2\alpha} \qquad\qquad \tan 3\alpha = \frac{3\tan\alpha - \tan^3\alpha}{1 - 3\tan^2\alpha}$$

The graph opposite is plotted in radians (π radians = 180°). A rough lookup table is also included here for emergency use.

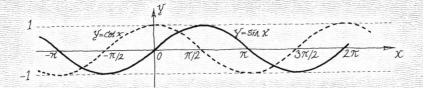

Radians	Degrees	Sine	Cosine	Tangent	Secant	Cosecant	Cotangent
0	0°	0	1	0	∞	1	∞
	2.5°	0.04362	0.9990	0.04366	22.926	1.00095	22.904
	5°	0.08716	0.9962	0.08749	11.474	1.00382	11.430
	7.5°	0.1305	0.9914	0.1317	7.6613	1.00863	7.5958
	10°	0.1736	0.9848	0.1763	5.7588	1.0154	5.6713
	12.5°	0.2164	0.9763	0.2217	4.6202	1.0243	4.5107
	15°	0.2588	0.9659	0.2679	3.8637	1.0353	3.7321
	17.5°	0.3007	0.9537	0.3153	3.3255	1.0485	3.1716
	20°	0.3420	0.9397	0.3640	2.9238	1.0642	2.7475
π/8	22.5°	0.3827	0.9239	0.4142	2.6131	1.0824	2.4142
	25°	0.4226	0.9063	0.4663	2.3662	1.1034	2.1445
	27.5°	0.4617	0.8870	0.5206	2.1657	1.1274	1.9210
	30°	0.5	0.8660	0.5774	2	1.1547	1.7321
	32.5°	0.5373	0.8434	0.6371	1.8612	1.1857	1.5697
	35°	0.5736	0.8192	0.7002	1.7434	1.2208	1.4281
	37.5°	0.6088	0.7934	0.7673	1.6427	1.2605	1.3032
	40°	0.6428	0.7660	0.8391	1.5557	1.3054	1.1918
	42.5°	0.6756	0.7373	0.9163	1.4802	1.3563	1.0913
π/4	45°	0.7071	0.7071	1	1.4142	1.4142	1
	47.5°	0.7373	0.6756	1.0913	1.3563	1.4802	0.9163
	50°	0.7660	0.6428	1.1918	1.3054	1.5557	0.8391
	52.5°	0.7934	0.6088	1.3032	1.2605	1.6427	0.7673
	55°	0.8192	0.5736	1.4281	1.2208	1.7434	0.7002
	57.5°	0.8434	0.5373	1.5697	1.1857	1.8612	0.6371
	60°	0.8660	0.5	1.7321	1.1547	2	0.5774
	62.5°	0.8870	0.4617	1.9210	1.1274	2.1657	0.5206
	65°	0.9063	0.4226	2.1445	1.1034	2.3662	0.4663
3π/8	67.5°	0.9239	0.3827	2.4142	1.0824	2.6131	0.4142
	70°	0.9397	0.3420	2.7475	1.0642	2.9238	0.3640
	72.5°	0.9537	0.3007	3.1716	1.0485	3.3255	0.3153
	75°	0.9659	0.2588	3.7321	1.0353	3.8637	0.2679
	77.5°	0.9763	0.2164	4.5107	1.0243	4.6202	0.2217
	80°	0.9848	0.1736	5.6713	1.0154	5.7588	0.1763
	82.5°	0.9914	0.1305	7.5958	1.00863	7.6613	0.1317
	85°	0.9962	0.08716	11.430	1.00382	11.474	0.08749
	87.5°	0.9990	0.04326	22.903	1.00095	22.926	0.04366
π/2	90°	1	0	∞	1	∞	1

SPHERICAL TRIGONOMETRY
formulæ for heaven and earth

A *spherical triangle* has internal angles whose sum is between 180° and 540°. Its sides are arcs of *great circles* (whose centers all lie at the sphere's center). Any two points on a sphere can define a *great circle*, and any three a *lesser circle*. Any circle on a sphere, greater or lesser, defines two *poles*.

The sides of a spherical triangle can be thought of as angles. The six relevant quantities are shown opposite and obey:

Law of Sines:
$$\frac{\sin a}{\sin A} = \frac{\sin b}{\sin B} = \frac{\sin c}{\sin C}$$

Law of Cosines:
$$\cos a = \cos b \cos c + \sin b \sin c \cos A$$
$$\cos A = -\cos B \cos C + \sin B \sin C \cos a$$

Law of Tangents:
$$\frac{\tan \frac{1}{2}(A+B)}{\tan \frac{1}{2}(A-B)} = \frac{\tan \frac{1}{2}(a+b)}{\tan \frac{1}{2}(a-b)}$$

Spherical trigonometry is used in navigation. For example, using degrees *longitude* and *latitude*, a ship sails from Q to R:

$$a = 90° - \text{lat.}\ R \qquad b = 90° - \text{lat.}\ Q \qquad C = \text{long.}\ Q - \text{long.}\ R$$

C is known as the *polar angle*. The initial and final courses are given by A and B and the length of the journey by c. Use the following equations to solve for B, A, and c:

$$\tan \frac{1}{2}(B+A) = \cos \frac{1}{2}(b-a)\ \sec \frac{1}{2}(b+a)\ \cot \frac{1}{2}C$$
$$\tan \frac{1}{2}(B-A) = \sin \frac{1}{2}(b-a)\ \csc \frac{1}{2}(b+a)\ \cot \frac{1}{2}C$$
$$\tan \frac{1}{2}c = \tan \frac{1}{2}(b-a)\ \sin \frac{1}{2}(B+A)\ \csc \frac{1}{2}(B-A)$$

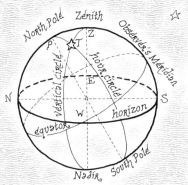

The Astronomical triangle PZT, for some celestial object

side TZ = zenith distance of T = 90° - altitude of T
side TP = polar distance of T = 90° - declination of T
side ZP = 90° - or + latitude of observer (N. or S. hemisphere)
angle PZT = azimuth of T (if T is east of observer's meridian)
   or 360° - azimuth of T (if T is west of OM)
angle ZPT = hour angle of T (if T is west of OM) in hours ' "
   or 360° - hour angle (if T is east of OM)

Celestial great circles

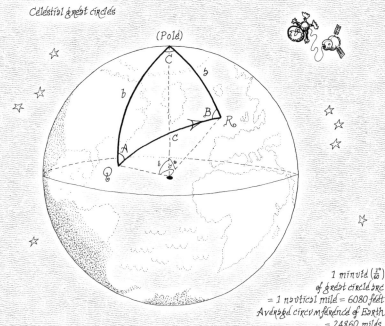

1 minute $\left(\frac{1°}{60}\right)$
of great circle arc
= 1 nautical mile = 6080 feet
Average circumference of Earth
= 24860 miles.

THE QUADRATIC FORMULA
discriminants and parabolas

A *quadratic equation* is one of the form $ax^2 + bx + c = 0$ (where a is not zero). The solutions, or *roots*, of such an equation are given by the quadratic formula,

$$\frac{-b \pm \sqrt{b^2 - 4ac}}{2a}$$

The quantity $b^2 - 4ac$ is called the *discriminant* which dictates the nature and number of the solutions. There are three cases:

$b^2 - 4ac > 0$ two different (real) solutions

$b^2 - 4ac = 0$ only one (real) solution

$b^2 - 4ac < 0$ two different *complex* or *imaginary* solutions
(as opposed to *real* ones—*see page 112*)

Examples (*shown opposite*):

i) $2x^2 - x - 1 = 0$ has a discriminant of 9 and produces two real solutions, 1 and $-\frac{1}{2}$.

ii) $x^2 - 2x + 1 = 0$ yields a discriminant of zero and hence has one real root, namely $x = 1$.

iii) $4x^2 + 8x + 5 = 0$ is an example of a quadratic equation with no real roots, solving for $x = -1 + \frac{i}{2}$ and $x = -1 - \frac{i}{2}$ (*see page 112*).

The quadratic equation $ax^2 + bx + c = 0$ has real solutions at values of x where the graph of the function $y = ax^2 + bx + c$ crosses the x-axis (*i.e.* where $y = 0$).

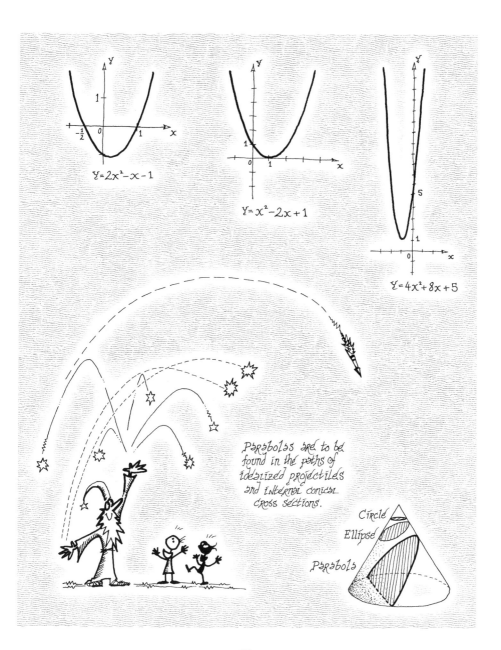

$Y = 2x^2 - x - 1$

$Y = x^2 - 2x + 1$

$Y = 4x^2 + 8x + 5$

Parabolas are to be found in the paths of idealized projectiles and internal conical cross sections.

Circle
Ellipse
Parabola

75

MATRICES AND VECTORS
basic operations and determinants

An $m \times n$ matrix is a rectangular array of numbers with m rows and n columns. Two $m \times n$ matrices can be added or subtracted in the obvious way, and any matrix can be similarly multiplied by a single number.

You can only multiply an $m \times n$ matrix by a $p \times q$ matrix if $n = p$ (i.e. the number of columns of the first equals the rows of the second). The procedure for matrix multiplication is shown below. Note that matrix multiplication is not commutative: AB is not necessarily equal to BA. In fact, BA is not defined here unless $m = q$. A *square* matrix has an equal number of rows and columns. The *determinant* $|A|$ of a square matrix A is an important scalar quantity associated with A (*see examples opposite*). Determinants of larger square matrices are defined iteratively.

The *inverse* of an $n \times n$ matrix A with $|A| \neq 0$ is an $n \times n$ matrix, A^{-1}, such that $AA^{-1} = A^{-1}A = I_n$ (the $n \times n$ identity matrix defined opposite).

Vectors describe *displacements* in space, e.g. $\mathbf{a} = (1, 1, 2)$. They are summed in the obvious way, with a closed circuit summing to a zero vector. The *dot* or *scalar product* of two vectors $\mathbf{a} \cdot \mathbf{b}$ is $\|\mathbf{a}\| \|\mathbf{b}\| \cos\theta$, where $\|\mathbf{a}\|$ is the length of the vector $\mathbf{a} = (x, y, z)$, e.g. $\sqrt{x^2 + y^2 + z^2}$, and θ is the angle between the vectors. The *cross product* of two three-dimensional vectors $\mathbf{a} \times \mathbf{b}$ is a vector perpendicular to both, $\|\mathbf{a}\| \|\mathbf{b}\| \sin\theta \, \mathbf{n}$, where $\mathbf{n}$ is the right-handed unit vector perpendicular to both.

$$\begin{pmatrix} a & b & c \\ d & e & f \end{pmatrix} \begin{pmatrix} p & s \\ q & t \\ r & u \end{pmatrix} = \begin{pmatrix} (ap + bq + cr) & (as + bt + cu) \\ (dp + eq + fr) & (ds + et + fu) \end{pmatrix}$$

identity matrices

$$\begin{bmatrix} 1 & 0 \\ 0 & 1 \end{bmatrix} \quad \begin{bmatrix} 1 & 0 & 0 \\ 0 & 1 & 0 \\ 0 & 0 & 1 \end{bmatrix} \quad \begin{bmatrix} 1 & 0 & 0 & 0 \\ 0 & 1 & 0 & 0 \\ 0 & 0 & 1 & 0 \\ 0 & 0 & 0 & 1 \end{bmatrix}$$

etc.

the transpose of a matrix

$$\begin{bmatrix} a & b & c \\ d & e & f \end{bmatrix} \text{ is } \begin{bmatrix} a & d \\ b & e \\ c & f \end{bmatrix}$$

the determinant of matrix

$$\begin{vmatrix} a & c \\ b & d \end{vmatrix} = \text{area of}$$

is $ad - bc$

(c,d)

(a,b)

the determinant of matrix

$$\begin{vmatrix} a & d & g \\ b & e & h \\ c & f & i \end{vmatrix} = \text{volume of}$$

is

$aei + dhc + bfg$
$- ceg - bdi - fha$

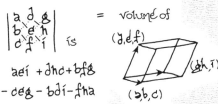

(d,e,f)

(g,h,i)

(a,b,c)

$$\begin{bmatrix} -1 & 0 \\ 0 & 1 \end{bmatrix}$$

reflection

$$\begin{bmatrix} 1 & 1.25 \\ 0 & 1 \end{bmatrix}$$

shearing

$$\begin{bmatrix} \frac{3}{2} & 0 \\ 0 & \frac{2}{3} \end{bmatrix}$$

scaling

$$\begin{bmatrix} \cos 30° & -\sin 30° \\ \sin 30° & \cos 30° \end{bmatrix}$$

rotation

EXPONENTIALS AND LOGARITHMS
growth and decay

Given some value a, we can define a *squared* and a *cubed* as follows: $a^2 = a \times a$, $a^3 = a \times a \times a$. In the expression a^n, n is the *exponent*. Here are the basic exponential formulæ:

$$a^0 = 1 \qquad\qquad a^p a^q = a^{p+q} \qquad\qquad (ab)^p = a^p b^p$$

$$a^{1/n} = \sqrt[n]{a} \qquad\qquad (a^p)^q = a^{pq} \qquad\qquad a^{m/n} = \sqrt[n]{a^m}$$

$$a^{-p} = \frac{1}{a^p} \qquad\qquad \sqrt[n]{\frac{a}{b}} = \frac{\sqrt[n]{a}}{\sqrt[n]{b}} \qquad\qquad \frac{a^p}{a^q} = a^{p-q}$$

The *base a logarithm* of x, $\log_a x = y$ is the quantity which satisfies $a^y = x$. Because $a^0 = 1$ and $a^1 = a$, then we always have both $\log_a a = 1$ and $\log_a 1 = 0$. Here are the essential logarithmic formulæ:

$$\log_a xy = \log_a x + \log_a y \qquad\qquad \log_a \frac{x}{y} = \log_a x - \log_a y$$

$$\log_a x^k = k \log_a x \qquad\qquad \log_a \frac{1}{x} = -\log_a x$$

$$\log_a \sqrt[n]{x} = \frac{1}{n} \log_a x \qquad\qquad \log_k a = \log_m a \, \log_k m$$

Any positive base (except 1) can be used, but most common are 10 and the constant e (= 2.718...), which occurs widely throughout the natural world, often in processes of growth and decay. $\log_e$ is usually just written *log* or *ln*. Logarithms allow multiplication and division of numbers by the addition or subtraction of exponents.

Exponentials are also useful for calculating *compound growth*. A quantity x which increases or decreases by p percent in unit time will, after time t, take the values $x\left(1+\frac{p}{100}\right)^t$ and $x\left(1-\frac{p}{100}\right)^t$ respectively.

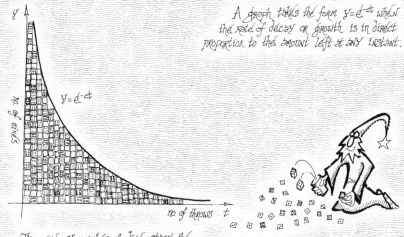

A graph takes the form $y = e^{-ct}$ when the rate of decay or growth is in direct proportion to the amount left at any instant.

$y = e^{-ct}$

no. of sixes

no. of throws t

Throwing y number of dice, stack the sixes. Rethrow the remainder, stack the sixes and repeat.

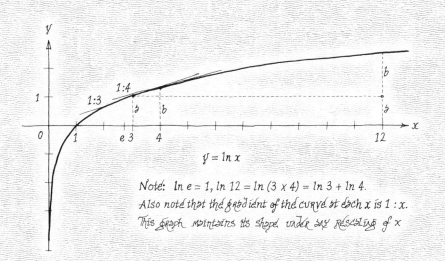

$y = \ln x$

Note: $\ln e = 1$, $\ln 12 = \ln (3 \times 4) = \ln 3 + \ln 4$.

Also note that the gradient of the curve at each x is $1 : x$.

This graph maintains its shape under any rescaling of x

MEANS AND PROBABILITIES
safe proportions and risky outcomes

Given two numbers, *a* and *b*, then the three important averages or *means,* traditionally used in geometry, music, and architecture, are:

the *arithmetic mean,* $\frac{1}{2}(a + b)$,   the *geometric mean,* $\sqrt{ab}$

and the *harmonic mean,* $\dfrac{2ab}{a+b}$

Suppose we have a situation with *n* possible, equally likely outcomes, *k* of these being desired. The probability *p* of a desired outcome occurring on any one occasion is then,

$$p = \frac{\textit{number of desired outcomes}}{\textit{total number of possible outcomes}} = \frac{k}{n}$$

Note that *p* is always between 0 and 1. Imagine *E* and *F* are two independent possible events, with probabilities $P(E)$ and $P(F)$ of occurring, respectively. The probability of both *E and F* occurring is $(EF) = P(E) \times P(F)$. With this, we can express the probability of *either E or F* occurring as $P(E + F) = P(E) + P(F) - P(EF)$.

F is *not* independent of *E* when the probability of *F* occurring is altered by *E* occurring. In this case we define the *conditional probability* $P(F|E)$ to be the probability of *F* occurring once *E* has occurred, and we have $P(EF) = P(E) \times P(F|E)$. For example, if G and H both represent the choice of a black ball from the bag on our left, then $P(GH) = \frac{2}{5} \times \frac{1}{4} = \frac{1}{10}$ (*see page 387 for fractions*).

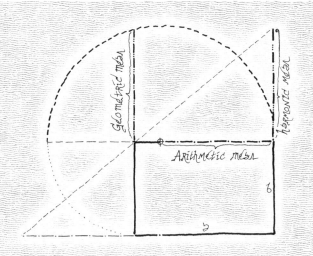

We are to choose one ball from each bag :

$P(E)$ = probability of choosing black ball from left bag = $\frac{2}{5}$

$P(F)$ = probability of choosing black ball from right bag = $\frac{4}{6}$ = $\frac{2}{3}$

$P(EF)$ = probability of choosing black balls from both bags = $\frac{2}{5} \times \frac{2}{3} = \frac{4}{15}$

$P(E+F)$ = probability of choosing at least one black ball = $\frac{2}{5} + \frac{2}{3} - \frac{4}{15} = \frac{4}{5}$

COMBINATIONS & PERMUTATIONS
ways of arranging things

Suppose we have n items and we wish to consider groupings of k of them. Two types of groupings exist: *combinations*, where the order is not important, and *permutations*, where the order does matter. We need to use *factorials* here. The factorial of m, written $m!$ (where $m \geq 1$) is defined as,

$$m! = m(m-1)(m-2) \ldots 2 \times 1 \quad (e.g.\ 6! = 6 \times 5 \times 4 \times 3 \times 2 \times 1 = 720)$$

0 is a special case: $0! = 1$ by convention.

The number of combinations of k items out of n is then,

$$C_k^n = \frac{n!}{(n-k)!\,k!}, \quad \text{also written } \binom{n}{k}$$

The (obviously larger) number of permutations is,

$$P_k^n = \frac{n!}{(n-k)!} = k!\binom{n}{k}$$

In a situation having two possible outcomes, P and Q, with probabilities p and q, we know $p+q = 1$, so $(p+q)^n = 1$. The term $\binom{n}{k}p^{n-k}q^k$ in the *binomial expansion* of $(p + q)^n$ is then the probability of $n-k$ occurrences of P and k occurrences of Q from a total of n. The general binomial formula is,

$$(x+y)^n = x^n + \binom{n}{1}x^{n-1}y + \ldots + \binom{n}{k}x^{n-k}y^k + \ldots + \binom{n}{n-1}xy^{n-1} + \binom{n}{n}y^n$$

The rows of *Pascal's Triangle* correspond here. For example:

$$(x+y)^4 = x^4 + 4x^3y + 6x^2y^2 + 4xy^3 + y^4$$

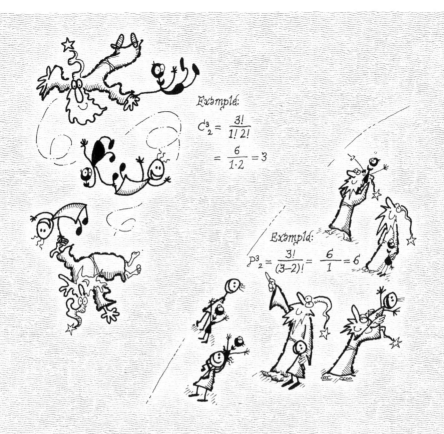

Example:

$$C^3_2 = \frac{3!}{1!\,2!}$$

$$= \frac{6}{1 \cdot 2} = 3$$

Example:

$$P^3_2 = \frac{3!}{(3-2)!} = \frac{6}{1} = 6$$

```
          1
        1   1
      1   2   1
    1   3   3   1
  1   4   6   4   1
 1  5  10  10  5   1
1  6  15  20  15  6   1
```

Pascal's Triangle – in which each number is the sum of the two numbers above it. Row $n+1$ corresponds to the terms in the expansion of $(x + y)^n$.

STATISTICS
distribution and deviation

Statistical analysis allows us to process samples of observed data in order to reveal trends and make predictions. If $x_1, x_2, \ldots, x_n$ is a set of values of some measurable phenomenon, the average or *mean* value is given by $\bar{x} = \frac{1}{n}(x_1 + \ldots + x_n)$.

The *standard deviation* σ of the sample then gauges the extent to which the data deviate from this mean:

$$\sigma = \sqrt{\frac{x_1^2 + \ldots + x_n^2}{n}} - \bar{x}^2 = \sqrt{\frac{\left(x_1^2 - \bar{x}^2\right) + \ldots + \left(x_n^2 - \bar{x}^2\right)}{n}}$$

The most commonly occurring form of *statistical distribution* is the *normal* or *Gaussian*. Its general form produces a bell-shaped curve centered at x whose 'width' is dependent on σ:

$$f(x) = \frac{1}{\sigma\sqrt{2\pi}}\, e^{-\frac{(x-x)^2}{2\sigma^2}}$$

In data which is *normally distributed* the probability of a value occurring in the range between a and b is equal to the area under the curve between these values as shown opposite. The total area under the curve (all possibilities) is equal to one.

The *Poisson distribution* tells us that if the mean number of events of a particular type in a fixed time interval is μ, then the probability of n events occurring in one interval is given by,

$$p(n) = \frac{\mu^n e^{-\mu}}{n!} \quad \text{where } e = 2.718\ldots \text{ (see page 78)}.$$

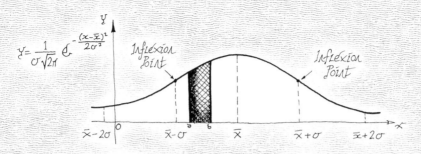

$$y = \frac{1}{\sigma\sqrt{2\pi}} e^{-\frac{(x-\bar{x})^2}{2\sigma^2}}$$

Inflexion Point

Inflexion Point

$\bar{x}-2\sigma$ $\bar{x}-\sigma$ σ $\bar{x}$ $\bar{x}+\sigma$ $\bar{x}+2\sigma$

The standard deviation σ becomes a convenient unit by which to scale the sample range. Note the positions of the inflective points.

68.26 % of sample lies within σ of $\bar{x}$.
95.44 % of sample lies within 2σ of $\bar{x}$.
99.73 % of sample lies within 3σ of $\bar{x}$.

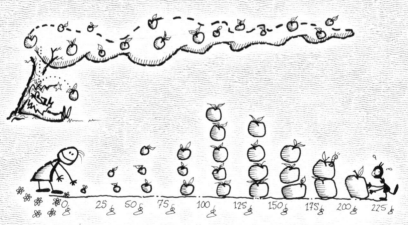

0g   25g   50g   75g   100g   125g   150g   175g   200g   225g

A random sample of apples arranged vertically according to a system of equal size brackets produces a frequency distribution graph. Given a large enough sample, such a graph would begin to approximate a continuous curve - the graph of a normal (or Gaussian) distribution.

KEPLER'S AND NEWTON'S LAWS
bodies in motion

Johannes Kepler (1571–1630) discovered three laws describing planetary motion. They hold true for all orbiting bodies in space.

1. *Planets move in ellipses with the sun at one focus.*

2. *A line drawn from the sun to the planet sweeps out equal areas in equal times.*

3. *The square of the time which a planet takes to orbit the Sun divided by the cube of its major semi-axis (see page 62) is a constant throughout the solar system.*

Using Kepler's discoveries, Isaac Newton (1643–1727) deduced his *Law of Universal Gravitation (see page 88)*, and then went on to deduce his general laws of motion:

1. *An object at rest or in motion will remain in such a state until acted upon by some force.*

2. *The acceleration produced by a force on a given mass is proportional to that force.*

3. *A force exerted by A on B is always accompanied by an equal force exerted by B on A, in the same straight line but in the opposite direction.*

Albert Einstein (1879–1955) discovered that at velocities approaching that of light, Newton's Laws require significant modification.

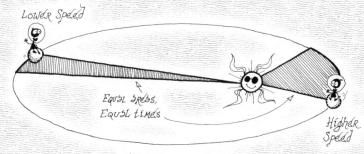

LOWER SPEED

Equal areas,
Equal times

Higher
Speed

Kepler's Law of Areas

Force = mass × acceleration

An object falling to Earth
exerts an equal and opposite
gravitational pull on the
Earth itself. However, this is
rarely acknowledged as the
Earth, being of much greater
mass, experiences no
noticeable acceleration.

GRAVITY AND PROJECTILES
featherless falling objects

Newton's *Law of Universal Gravitation* states that any two masses m_1 and m_2 at a distance d apart will exert equal and opposite forces on each other of magnitude F where,

$$F = \frac{Gm_1 m_2}{d^2}$$

G is the *universal gravitation constant (see page 384)*. Note that d is the distance between the *centers of mass* of m_1 and m_2.

Near the surface of the Earth (mass m_2), d is effectively constant, giving the 'local' gravitational *acceleration* constant g:

$$g = \frac{Gm_2}{d^2} = 9.80665 \text{ m/sec}^2 \text{ so that } F = m_1 g$$

Assuming negligible air resistance, an object falling from rest will travel s meters in t seconds where s and t are related by,

$$s = \tfrac{1}{2}gt^2 \quad \text{or} \quad t = \sqrt{\frac{2s}{g}}$$

The object's velocity at time t will be $v = gt = \sqrt{2gs}$ m/sec. Note that these quantities are independent of mass.

The path of a projectile with initial velocity v and angle of trajectory θ is given by,

$$x(t) = vt\cos\theta \quad \text{and} \quad y(t) = vt\sin\theta - \tfrac{1}{2}gt^2$$

These are time-dependent coordinates.

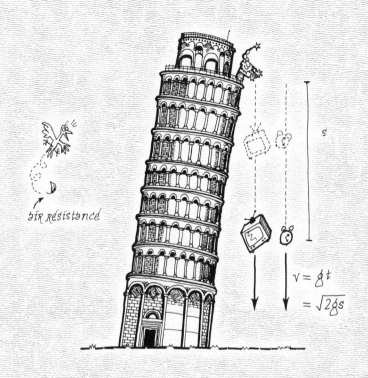

air resistance

s

$v = gt$

$= \sqrt{2gs}$

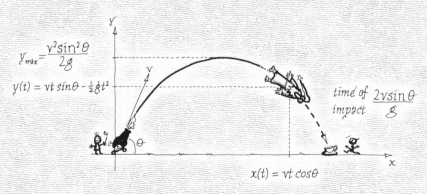

$y_{max} = \dfrac{v^2 \sin^2 \theta}{2g}$

$y(t) = vt \sin\theta - \frac{1}{2}gt^2$

$\text{time of impact} \quad \dfrac{2v\sin\theta}{g}$

$x(t) = vt \cos\theta$

ENERGY, WORK, AND MOMENTUM
conservation in action

An object, mass m, moving in a straight line with velocity v has *kinetic energy* $E_k = \frac{1}{2}mv^2$. This is energy it possesses because of its motion. When an applied force changes its velocity to u, the total *work* done, W, is the change in kinetic energy: $W = \frac{1}{2}mv^2 - \frac{1}{2}mu^2$

Generally, *work* measures an exchange of energy between two bodies. Someone lifting an object of mass m to a height h above its initial position does work, here transferring gravitational *potential energy*, E_p, to the object (it can now fall). $E_p = mgh$ (mg is the *weight* of the object, a force).

When the object falls, it loses height but gains velocity, thus E_k increases as E_p decreases. Ignoring friction, the total energy of the object $E = E_k + E_p$ remains constant until it lands, when its remaining kinetic energy is dissipated as heat and noise.

The *linear momentum* of an object is given by $p = mv$. For a point mass m rotating about an axis at distance r, the *angular momentum*, L equals $(mv)r = (m\omega r)r = mr^2\omega$ where ω is the *angular velocity* of the body, in radians per second. $I = mr^2$ is known as the *moment of inertia*. The *rotational kinetic energy* of a system is then $E_{kr} = \frac{1}{2}I\omega^2$.

A general rotating solid can be treated as if it were a point mass rotating about the same axis with the appropriate *radius of gyration*. This is shown opposite and can be found using the methods of calculus (*see page 110*). If no external forces act on a system, its total momentum is always conserved.

Kinetic energy is conserved so
$$\tfrac{1}{2}MV^2 = \tfrac{1}{2}Mp^2 + \tfrac{1}{2}mq^2$$
also because linear momentum is conserved:
$$Mp\sin\alpha - mq\sin\beta = 0$$
(horizontal component)
$$Mp\cos\alpha + mq\cos\beta = MV$$
(vertical component)

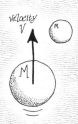

velocity V

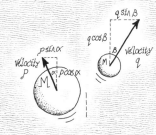

$q\sin\beta$

$q\cos\beta$

velocity q

$p\sin\alpha$

velocity p

$p\cos\alpha$

force = Rate of change of Momentum

The moment of inertia, I of a body rotating about an axis is given as

$$I = \sum Mr^2 = \int r^2 \, dM$$

Torque results in angular acceleration α such that $T = I\alpha$

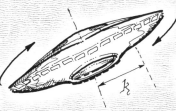

$I = MK^2$ where M is the total mass & K is the radius of gyration.

ROTATION AND BALANCE
whirling, gears, and pulleys

If an object of mass m on a string of length r is swung around in a circle at a velocity of v, then there is a *centripetal force*:

$$F = \frac{mv^2}{r} = m\omega^2 r \quad \text{giving acceleration} \quad a = \frac{v^2}{r} = \omega^2 r$$

toward the center, where ω is the angular velocity. The centripetal force is equal and opposite to the string tension, which is also treated as a force.

Two interlocking gears with t_1 and t_2 teeth, and speeds r_1 and r_2 (in rpm or any other unit), relate by,

$$t_1 \, r_1 = t_2 \, r_2 \quad \text{or equivalently,} \quad r_1 = \frac{t_2}{t_1} \, r_2 \text{ and } r_2 = \frac{t_1}{t_2} \, r_1$$

The equation also holds true for two belt pulley wheels with diameters t_1 and t_2, and speeds r_1 and r_2.

If two objects with weights w_1 and w_2 are balanced, as shown opposite, at distances d_1 and d_2 from a fulcrum, then the *torque* associated with the two objects must be equal. Torque is the product of force and radial distance:

$$d_1 \, w_1 = d_2 \, w_2$$

A long-handled wrench turns a nut more easily than a short-handled one because it produces more torque.

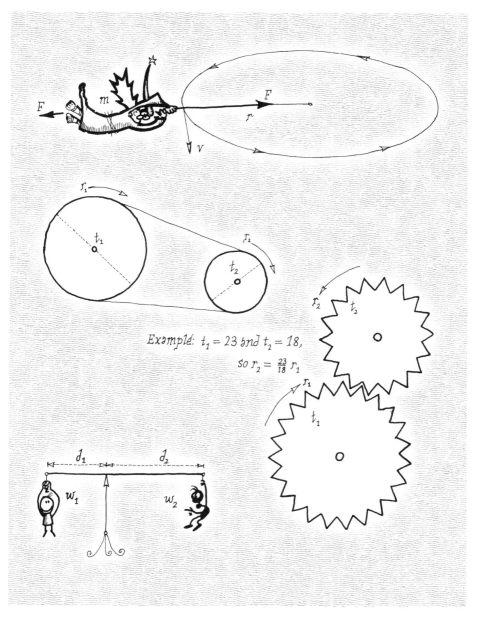

Example: $t_1 = 23$ and $t_2 = 18$,

so $r_2 = \frac{23}{18} r_1$

SIMPLE HARMONIC MOTION
vibrating and oscillating phenomena

———————

Galileo Galilei (1564–1642) discovered that the *period T* of a pendulum, the time it takes to swing from one side to the other and back again, is independent of its *amplitude*, its maximum displacement from the center. Thus a pendulum of length *l* completes *f* swings every second, regardless of whether its swings are wide or narrow. Here *f* is the *frequency* of the pendulum.

If *l* is given in meters, the period *T* is given in seconds by,

$$T = 2\pi\sqrt{\frac{l}{g}} = \frac{1}{f}$$

where *g* is the gravitational constant (*see page 384*).

For small swings, less than 5°, pendula approximate *simple harmonic motion*. Bobbing bottles and sproinging springs all perform *s.h.m.*, as do a vast array of vibrating and oscillating phenomena. Potential and kinetic energies are continually exchanged. The potential energy and acceleration of the vibrating object reach maxima when its kinetic energy and velocity are minimal and vice versa.

A simple way to generate *s.h.m.* is to project a uniform circular motion onto an axis as shown opposite, giving,

$$d(t) = a \sin \omega t$$

where *a* is the amplitude and *ω* the angular velocity. The period is then $\frac{2\pi}{\omega}$ seconds per cycle and the frequency is the reciprocal of the period, $\frac{\omega}{2\pi}$ cycles per second.

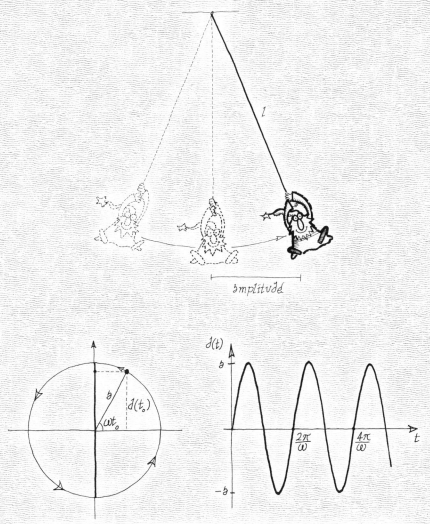

Projecting the uniform circular motion onto the vertical axis results in a non-uniform up-and-down motion directly related to the sine function.

STRESS, STRAIN, AND HEAT

expansion, contraction, tension, and compression

When a material is stretched or squashed it changes shape. The *stress*, σ, in the material is defined as the force F per unit area A; the *strain*, ε, is defined as the change in length, Δl, relative to the original length, l_0. *Young's Modulus E* for the material is then,

$$E = \frac{stress}{strain} = \frac{\sigma}{\varepsilon} = \frac{F/A}{\Delta l/l_0}$$

Substances also possess a *Bulk Modulus K*, relating inversely to its volumetric *compressibility*, and a *Shear Modulus G*, the ratio of shear stress and strain (*shown opposite*).

Heating and cooling cause expansion and contraction of materials in a linear relation to the temperature increase or decrease. The *linear expansivity*, α, of a substance relates its change in length Δl to its change in temperature ΔT, so that $\Delta l = \alpha l_0 \Delta T$, where l_0 is its starting length.

Hooke's Law states that a spring (or equivalently elastic object) stretched x units of distance beyond its equilibrium position pulls back with a force $F = kx$, where k is the *spring constant* of the spring in question. As a consequence, when a weight is attached to a vertical spring, the elongation is proportional to the weight.

For any structure to be in equilibrium, the total forces at any point must be balanced as shown opposite.

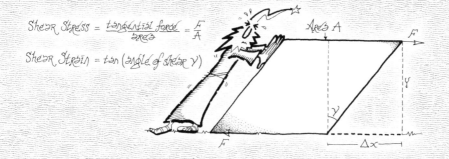

Shear Stress = $\dfrac{\text{tangential force}}{\text{area}} = \dfrac{F}{A}$

Shear Strain = tan (angle of shear γ)

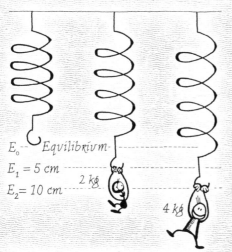

E_0 — — Equilibrium - - - - - - -

$E_1 = 5$ cm

$E_2 = 10$ cm   2 kg

4 kg

If a 2 kg. force stretches a spring 5 cm,
we can expect a 4 kg force to stretch it 10 cm.
This law applies up to a certain point
called the elastic limit.

At each point a
balanced triangle
of forces exists

LIQUIDS AND GASES
temperature, pressure, and flow

Liquid flowing at velocity v though a pipe with a cross-sectional area A has a rate of flow q where $q = Av$.

Liquid flowing through two pipes of cross-sectional areas A_1 and A_2, both subjected to the same pressure, will have flow velocities v_1 and v_2, where $A_1 v_1 = A_2 v_2$.

Pascal's Principle states that pressure applied to an enclosed liquid of any shape is transmitted evenly throughout, where pressure is defined as force per unit area. In the example illustrated (*opposite top*), $F_1 A_2 = F_2 A_1$.

Bernoulli's Equation (*illustrated lower opposite*) states that a change of height results in a pressure change in a liquid: $p_1 + h_1 \rho g = p_2 + h_2 \rho g$, where ρ is the average density of the fluid.

Turning to gases, the *Perfect Gas Law* states that if a fixed quantity of gas has pressure P, volume V, and temperature T (in degrees *Kelvin*), then PV is proportional to T. The Kelvin scale is a measure of *absolute temperature*, where $°K = °C + 273.15$.

Given a closed system with initial pressure, volume, and temperature P_1, V_1, and T_1, and later values P_2, V_2, and T_2:

Charles' Law states: $\dfrac{V_1}{V_2} = \dfrac{T_1}{T_2}$ for a constant pressure,

Boyle's Law states: $P_1 V_1 = P_2 V_2$ for a constant temperature.

Pressure applied to an enclosed liquid of any shape is transmitted evenly throughout. This is Pascal's principle, the basis of hydraulics.

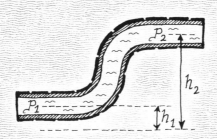

Here ρ denotes the average density of the fluid, given by mass/volume. If we are using the metric unit g/cm^3, then water gives $\rho = 1$. g is the gravitational constant

99

SOUND
harmonic wavelengths and passing sirens

For a string fixed at two points distance L apart, as shown, the harmonic wavelengths which fit the string are given by,

$$\lambda_n = \frac{2L}{n} \text{ for } n = 1, 2, 3 \ldots$$

Only λ_1 and the first few *overtones* may actually be audible.

If W_T is the wave velocity in the string, T its tension (in newtons), and μ the mass per unit length (in kg/m), then the basic frequencies, v_n, (in cycles per second) are,

$$v_n = \frac{W_T}{\lambda_n} = \frac{n}{2L} W_T = \frac{n}{2L} \sqrt{\frac{T}{\mu}}$$

The tone of an ambulance siren appears to change as it passes. This is known as the *Döppler Effect*. If an observer is moving at velocity v_o *toward* a source of frequency f_s, which is itself moving at velocity v_s *toward* the observer, then for a wave of speed c the observed frequency f_o is,

$$f_o = \left(\frac{c + v_o}{c - v_s} \right) f_s$$

Note: if objects recede rather than approach, use negative values for v_o and v_s (the speed of sound is 331.45 m/sec).

If two very similar tones, frequencies f_1 and f_2, are played together, a *beat frequency* $f_{beat} = (f_2 - f_1)$ may be heard.

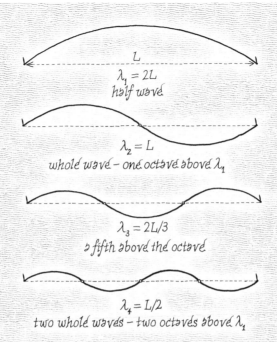

$\lambda_1 = 2L$
half wave

$\lambda_2 = L$
whole wave – one octave above λ_1

$\lambda_3 = 2L/3$
a fifth above the octave

$\lambda_4 = L/2$
two whole waves – two octaves above λ_1

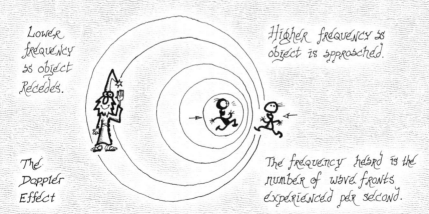

Lower frequency as object recedes.

Higher frequency as object is approached.

The Doppler Effect

The frequency heard is the number of wave fronts experienced per second.

LIGHT
refraction, lenses, and relativity

Light travelling from air into water is shown opposite. If v_1 and v_2 are the speeds that light takes in any two media then *Snell's Law of Refraction* states that for a given frequency:

$$\frac{v_1}{v_2} = \frac{\sin\theta_1}{\sin\theta_2} = \text{constant} \quad \text{or} \quad n_1 \sin\theta_1 = n_2 \sin\theta_2$$

where n_1 and n_2 are the *refractive indices* of the two media (varying slightly for different frequencies). Here are a few useful values:

Vacuum and air: 1   *Water*: 1.33   *Quartz*: 1.45   *Crown glass*: 1.52

Various lenses are shown opposite, some converging, others diverging. The *focal length, f,* is shown for one. The *Gaussian Lens Equation* relates the distances between an object, the lens, and the upside-down focused image:

$$\frac{1}{x_o} + \frac{1}{x_i} = \left(\frac{n_{lens}}{n_{medium}} - 1\right)\left(\frac{1}{R_1} + \frac{1}{R_2}\right) = \frac{1}{f}$$

Here R_1 and R_2 are the left and right *radii of curvature* of the lens (negative if concave).

Visible light is a tiny part of the *electromagnetic spectrum* which also includes x-rays, radio, and microwaves. Einstein was able to deduce that, because light travels at a constant speed away from you irrespective of your own speed, time itself must be able to stretch and dilate! This is part of his *Special Theory of Relativity.*

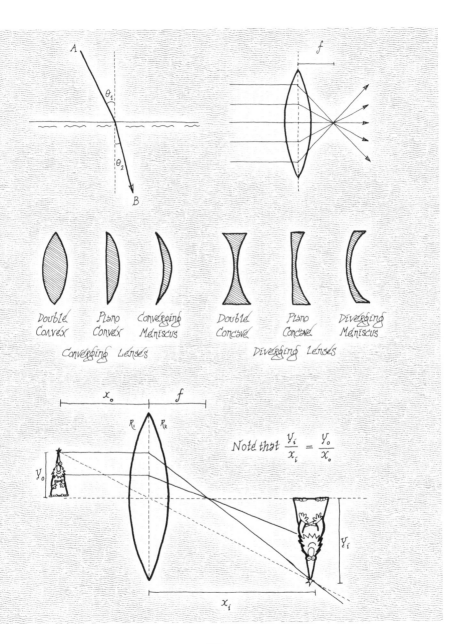

A

θ_1

θ_2

B

f

Double
Convex

Plano
Convex

Converging
Meniscus

Double
Concave

Plano
Concave

Diverging
Meniscus

Converging Lenses

Diverging Lenses

x_o

f

R_1 R_2

Note that $\dfrac{y_i}{x_i} = \dfrac{y_o}{x_o}$

y_o

y_i

x_i

103

ELECTRICITY AND CHARGE
tuning the circuit

For a simple electrical circuit, the *voltage E* (in volts) across a *resistance R* (in ohms, Ω) produces a *current I* (in amps) defined by *Ohm's Law*, $E = IR$. The *power P* (in watts) in the circuit is then,

$$P = EI = I^2 R$$

Resistors in series give resistance $R_s = R_1 + \ldots + R_n$ ohms. Capacitors in parallel give capacitance $C_p = C_1 + \ldots + C_n$ faradays. Resistors in parallel and capacitors in series combine to give,

$$R_p = \cfrac{1}{\cfrac{1}{R_1} + \ldots + \cfrac{1}{R_n}} \qquad C_s = \cfrac{1}{\cfrac{1}{C_1} + \ldots + \cfrac{1}{C_n}}$$

Equations for circuits involving inductors are shown opposite.

All electrical effects are the result of *charge* (measured in coulombs). The charge of an electron is -1.6×10^{-19} C. *Coulomb's Law* states that the force F between two point charges Q_1 and Q_2 separated by a distance of r meters is given by,

$$F = \frac{Q_1 Q_2}{4\pi\varepsilon_0 r^2}$$

where ε_0 is the *permittivity of empty space*, 8.85×10^{-12} farad/m. In the microcosm, Coulomb forces bind electrons to nuclei to form atoms, atoms to atoms to form molecules, and molecules to molecules to form solids and liquids.

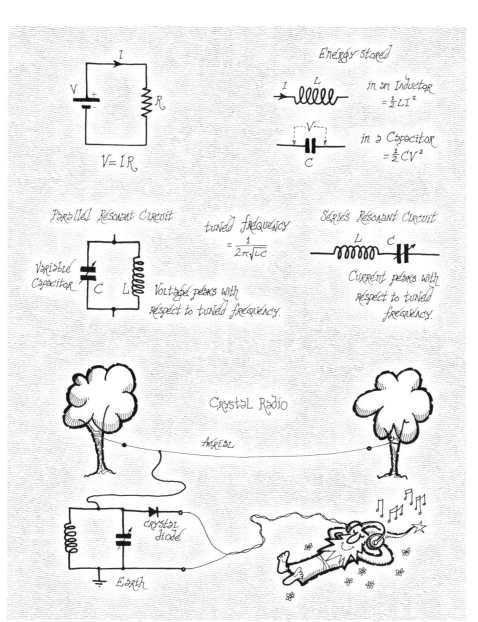

$V = IR$

Energy stored

in an Inductor
$= \frac{1}{2} L I^2$

in a Capacitor
$= \frac{1}{2} C V^2$

Parallel Resonant Circuit

Variable Capacitor

Voltage peaks with respect to tuned frequency.

tuned frequency
$= \frac{1}{2\pi\sqrt{LC}}$

Series Resonant Circuit

Current peaks with respect to tuned frequency.

Crystal Radio

Aerial

crystal diode

Earth

ELECTROMAGNETIC FIELDS
charge, flux, and handedness

Charge behaves in an electromagnetic field similarly to mass in a gravitational field. The force **F** on a moving charge Q in an electromagnetic field of strength **E** is given by **F** = EQ. The force **F** on a wire carrying a current I and with length l is **F** = **B**$I$$l$, where **B** is the *magnetic flux density*, measured in *teslas*. Other situations are shown opposite (μ_0 is the permeability of free space, the magnetic constant). Note: bold-faced quantities are vectors, with directions as well as magnitude.

A point charge Q moving with velocity v creates a magnetic field **B** (*opposite top left*). If point P has distance r from Q and is at angle θ to its direction of motion, then the *Biot-Savart Law* states that

$$\mathbf{B} = \left(\frac{\mu_0}{4\pi}\right)\frac{Qv\sin\theta}{r^2}$$

A moving magnetic field produces an electric field and vice versa. Using magnets and currents in coils of wire electrical energy can be transformed into mechanical energy (a *motor*) and vice versa (a *dynamo*). Fleming's left- and right-hand rules apply respectively (*lower opposite*).

Often useful in designing such things are *Faraday's Law of Induction*, (one of Maxwell's four equations) which states that the induced electromotive force in any closed circuit is proportional to the rate of change of the magnetic flux, or $\nabla\times \mathbf{E} = -\frac{\partial \mathbf{B}}{\partial t}$, where $\nabla\times$ is the *curl operator*, and *Lenz's Law*, which states that an induced current is always in such a direction as to oppose the motion or change causing it.

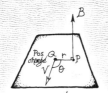

current is in direction of
movement of positive charge.

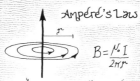

Ampéré's Law

$$B = \frac{\mu_0 I}{2\pi r}$$

field strength at distance
r from a wire.

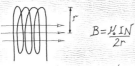

$$B = \frac{\mu_0 IN}{2r}$$

for a coil radius r with N turns

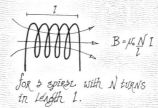

$$B = \mu_0 \frac{N I}{l}$$

for a spiral with N turns
in length L.

PULSED D.C. DYNAMO (WATER POWERED)

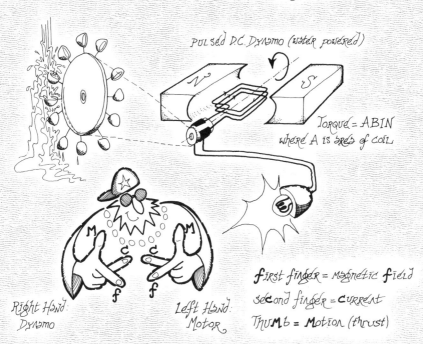

$$Torque = ABIN$$

where A is area of coil

Right Hand:
Dynamo

Left Hand:
Motor

first finger = Magnetic field
second finger = current
THUMB = Motion (thrust)

CIRCUITS AND LOGIC GATES
chips with everything

A *diode* acts as a one-way valve. Power supplies often need to change AC (*alternating current*) into DC (*direct current*). A simple single diode rectifier does only half the job (*opposite top*). A more elaborate bridge is better, allowing both parts of the AC cycle through. Add a capacitor to smooth things over and presto—a useful AC to DC converter!

Transistors are more like a tap or faucet, where a voltage at the base (B) controls the current flow from the collector (C) to the emitter (E). They can then be used as switches or amplifiers (*center opposite*). Note that since the output here depends on how much of the supply *doesn't* get bypassed through the transistor, it has inverted phase to the input.

Etching transistors and other components onto silicon chips enables heaps of circuitry to be crammed into a tiny space. The high gain operational amplifier, and the demultiplexer, used in data transmission, are fairly basic integrated circuits (*lower opposite*). A fast computer chip can be home to billions of transistor switches, linked together to form *logic gates* based on the binary mathematics of *Boolean operators* (*below*).

input		NOT	AND	OR	NAND	NOR	XOR	XNOR
A → B →								
A	B	$\overline{A}$	A·B	A+B	$\overline{A \cdot B}$	$\overline{A+B}$	A⊕B	$\overline{A \oplus B}$
0	0	1	0	0	1	1	0	1
0	1	1	0	1	1	0	1	0
1	0	0	0	1	1	0	1	0
1	1	0	1	1	0	0	0	1

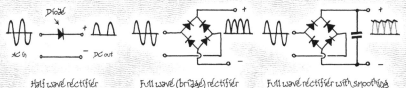

Half wave rectifier   Full wave (bridge) rectifier   Full wave rectifier with smoothing

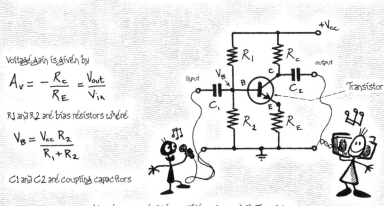

Voltage gain is given by

$$A_v = -\frac{R_c}{R_E} = \frac{V_{out}}{V_{in}}$$

R1 and R2 are bias resistors where

$$V_B = \frac{V_{cc}\,R_2}{R_1 + R_2}$$

C1 and C2 are coupling capacitors

Simple common emitter amplifier using an NPN Transistor

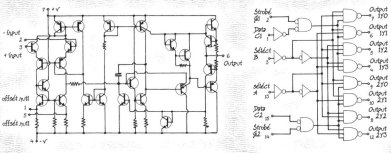

741 Op-Amp (Operational Amplifier) circuit schematic   Inside a 74155 demultiplexer chip

CALCULUS
differentiation and integration

Calculus makes use of *infinitesimals* and *limits* to solve two problems, the instantaneous rate of change of a function and the exact area under a curve.

The graph of the function $y = f(x)$ at position $(a, f(a))$ has gradient $f'(a)$, where the function $f'(x) = \frac{df}{dx}(x)$ is called the *derivative* of f. For each $x, f'(x)$ is the rate of change of f at x. To directly calculate such a derivative at a, we consider the slopes of the lines through $(a, f(a))$ and $(a+\varepsilon, f(a+\varepsilon))$ for ever-tinier values of ε. If they tend toward a 'limit', then the rate of change of f at a can be defined to be this limit.

If $x(t)$ denotes the position of an object at time t, then its velocity $v(t)$ at time t is $x'(t) = \frac{dx}{dt}(t)$. Its acceleration $a(t)$, the rate of change of its velocity at t, is then $x''(t) = \frac{dv}{dt}(t) = \frac{d^2x}{dt^2}(t)$.

Suppose we have a function (*opposite*) and seek the area beneath its graph between a and b. The interval between a and b is divided into an ever-larger number of equal lengths. This produces an ever-larger number of narrowing rectangles, the sum of whose areas can easily be found at each stage. The area under the curve is given by the limit of these sums, written $\int_a^b f(x)dx$. If $F(x)$ satisfies $F'(x) = f(x)$, then remarkably, $\int_a^b f(x)dx = F(b) - F(a)$.

$F(x)$ is called the *indefinite integral* or *antiderivative* of f and denoted $\int f\, dx$. As $(F(x) + c)' = F'(x)$, it follows that all of the antiderivatives given opposite involve an arbitrary constant.

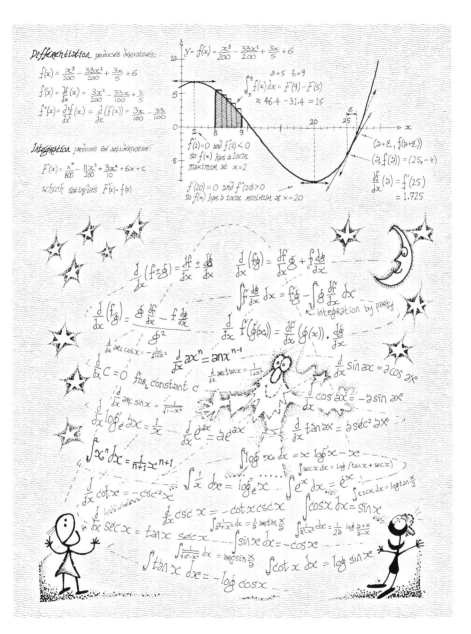

Differentiation produces derivatives:

$$f(x) = \frac{x^3}{200} - \frac{33x^2}{200} + \frac{3x}{5} + 6$$

$$f'(x) = \frac{df}{dx}(x) = \frac{3x^2}{200} - \frac{33x}{100} + \frac{3}{5}$$

$$f''(x) = \frac{d^2f}{dx^2}(x) = \frac{d}{dx}\left(f'(x)\right) = \frac{3x}{100} - \frac{33}{100}$$

Integration produces the antiderivative:

$$F(x) = \frac{x^4}{800} - \frac{11x^3}{200} + \frac{3x^2}{10} + 6x + c$$

which satisfies $F'(x) = f(x)$

$$y = f(x) = \frac{x^3}{200} - \frac{33x^2}{200} + \frac{3x}{5} + 6$$

$$a = 5 \quad b = 9$$

$$\int_5^9 f(x)\,dx = F(9) - F(5)$$
$$\approx 46.4 - 31.4 = 15$$

$f'(2) = 0$ and $f''(2) < 0$
so $f(x)$ has a local maximum at $x = 2$

$f'(20) = 0$ and $f''(20) > 0$
so $f(x)$ has a local minimum at $x = 20$

$$(a+\varepsilon, f(a+\varepsilon))$$
$$(a, f(a)) = (25, -4)$$
$$\frac{df}{dx}(a) = f'(25)$$
$$= 1.725$$

$$\frac{d}{dx}(f \pm g) = \frac{df}{dx} \pm \frac{dg}{dx} \qquad \frac{d}{dx}(fg) = \frac{df}{dx}\,g + f\frac{dg}{dx}$$

$$\int f\frac{dg}{dx}\,dx = fg - \int g\frac{df}{dx}\,dx \quad \leftarrow \text{integration by parts}$$

$$\frac{d}{dx}\left(\frac{f}{g}\right) = \frac{g\frac{df}{dx} - f\frac{dg}{dx}}{g^2} \qquad \frac{d}{dx}f(g(x)) = \frac{df}{dx}(g(x))\cdot\frac{dg}{dx}$$

$$\frac{d}{dx}\arccos x = -\frac{1}{\sqrt{1+x^2}} \qquad \frac{d}{dx}ax^n = anx^{n-1} \qquad \frac{d}{dx}\sin ax = a\cos ax$$

$$\frac{d}{dx}\arctan x = \frac{1}{1+x^2}$$

$$\frac{d}{dx}C = 0 \text{ for constant } c \qquad \frac{d}{dx}\cos ax = -a\sin ax$$

$$\frac{d}{dx}\arcsin x = \frac{1}{\sqrt{1-x^2}}$$

$$\frac{d}{dx}\log_e x = \frac{1}{x} \qquad \frac{d}{dx}e^{ax} = ae^{ax} \qquad \frac{d}{dx}\tan ax = a\sec^2 ax$$

$$\int x^n\,dx = \frac{1}{n+1}x^{n+1} \qquad \int \log_e x\,dx = x\log_e x - x \qquad \int \sec x\,dx = \log(\tan x + \sec x)$$

$$\frac{d}{dx}\cot x = -\csc^2 x$$

$$\int \frac{1}{x}\,dx = \log_e x \qquad \int e^x\,dx = e^x \qquad \int \csc x\,dx = \log\tan\frac{x}{2}$$

$$\frac{d}{dx}\csc x = -\cot x\,\csc x \qquad \int \cos x\,dx = \sin x$$

$$\frac{d}{dx}\sec x = \tan x\,\sec x \qquad \int\frac{1}{a^2+x^2}dx = \frac{1}{a}\arctan\frac{x}{a} \qquad \int\frac{1}{a^2-x^2}dx = \frac{1}{2a}\log\frac{a+x}{a-x}$$

$$\int \sin x\,dx = -\cos x$$

$$\int\frac{1}{\sqrt{a^2-x^2}}dx = \arcsin\frac{x}{a} \qquad \int \cot x\,dx = \log\sin x$$

$$\int \tan x\,dx = -\log\cos x$$

COMPLEX NUMBERS
into the imaginary realm

The familiar *real numbers* are contained within the larger realm of *complex* numbers. These are constructed by starting with the *imaginary unit* which is denoted i, and which satisfies (unlike any real number):

$$i^2 = -1 \text{ or } i = \sqrt{-1}$$

Given any two real numbers a and b, the quantity $a + bi$ is called a complex number. Some complex equations are shown below:

$$a + bi = c + di \text{ if and only if } a = c \text{ and } b = d$$

$$(a + bi) + (c + di) = (a + c) + (b + d)i$$

$$(a + bi)(c + di) = (ac - bd) + (ad + bc)i$$

$$\frac{ac + bd}{c^2 + d^2} = \frac{a + bi}{c + di} + \frac{bc - ad}{c^2 + d^2}i$$

Polar representation uses the angle θ and radius r:

$$z = r\cos\theta + ir\sin\theta = r(\cos\theta + i\sin\theta)$$

The exponential function e^x may be extended into the complex plane using *Euler's equation,*

$$e^{i\theta} = \cos\theta + i\sin\theta$$

From this we have both the mathematical gem $e^{i\pi} = -1$, and *DeMoivre's Theorem,* for powers of a complex number z:

$$z^n = (re^{i\theta})^n = r^n e^{in\theta} = r^n(\cos n\theta + i\sin n\theta)$$

-5 -4 -3 -2 -1 0 1 2 3 4 5

The real number line contains not only the positive and negative whole numbers, but all positive and negative fractions and irrational numbers such as $\sqrt{2}$ and π.

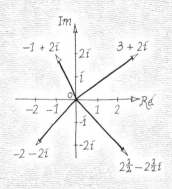

The real number line becomes the Re(al) axis in the plane of complex numbers. Each point in the plane corresponds to a complex number, and vice versa. Numbers on the Im(aginary) axis are called 'pure imaginary', having a real component of 0.

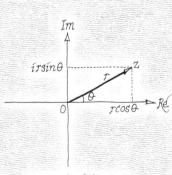

'polar' representation of a complex number

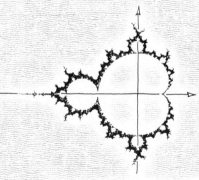

The complex plane is the home of the famed fractal Mandelbrot Set, which can be generated using a process involving the iteration of the map $z \to z^2 + c$.

Relativity Theory
length contractions and time dilations

Einstein extended Galileo's principle of relativity of uniform motion from mechanics to all laws of physics (including electrodynamics), while incorporating the constant nature of the speed of light, c.

At the core of his *Special Theory of Relativity* are the *Lorentz transformations* relating a coordinate frame (x', y', z', t') to another, (x, y, z, t), which it is moving relative to at velocity v. In the special case where motion is parallel to (for example) the x-axis, these are:

$$x' = \gamma(x - vt), \; y' = y, \; z' = z, \; t' = \gamma(t - vx/c^2) \text{ where } \gamma = \left(1 - v^2/c^2\right)^{-1/2}$$

Note that these transformations result in length contraction in the x-direction *and* time dilation (but both are negligible when v is small relative to c). For a more general direction of motion, a matrix-based equation provides the analogous transformations.

Composing velocities u and v in this context gives not $u + v$, but,

$$\frac{u + v}{1 + \left(\frac{uv}{c^2}\right)}$$

The (scalar) energy and the momentum (vector) of an object with mass m and velocity (vector) v are given, respectively, by

$$E = \gamma mc^2, \; p = \gamma mv$$

and if the object is at rest, we get that most famous of equations,

$$E = mc^2$$

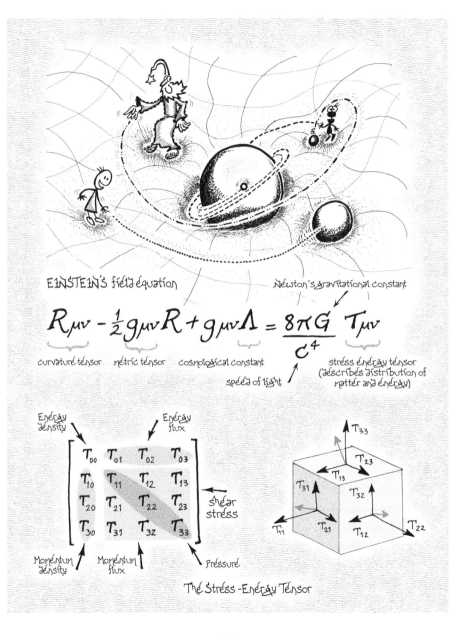

EINSTEIN'S field equation

Newton's gravitational constant

$$R_{\mu\nu} - \frac{1}{2} g_{\mu\nu} R + g_{\mu\nu} \Lambda = \frac{8 \pi G}{c^4} T_{\mu\nu}$$

curvature tensor   metric tensor   cosmological constant

speed of light

stress energy tensor
(describes distribution of
matter and energy)

Energy density   Energy flux

$$\begin{bmatrix} T_{00} & T_{01} & T_{02} & T_{03} \\ T_{10} & T_{11} & T_{12} & T_{13} \\ T_{20} & T_{21} & T_{22} & T_{23} \\ T_{30} & T_{31} & T_{32} & T_{33} \end{bmatrix}$$

shear stress

Momentum density   Momentum flux   Pressure

The Stress-Energy Tensor

QUANTUM MECHANICS
in a nutshell

Quantum mechanics is founded on *Planck's postulate*, which states that electromagnetic energy can only be emitted as multiples of some elementary unit, depending on the frequency of radiation v. This elementary unit or *quantum E* is given by the *Planck relation*:

$$E = hv$$

where h is *Planck's constant*, $6.2606896 \times 10^{-34}$ J·s.

Heisenberg's uncertainty principle tells us that the more precisely you determine the location x of a particle, the less you can say about its momentum p (and vice versa). In its simplest form,

$$\Delta x \Delta p \geq \frac{\hbar}{2} \ \ (\textit{Dirac's constant } \hbar = \frac{h}{2\pi})$$

where Δ relates to the amount of uncertainty in the value of x or p.

The *quantum state* of a physical system is described by a *wave function* Ψ which 'encodes' the probabilities of the various configurations of this system occurring at any given time. Its evolution is governed by *Schrödinger's equation*:

$$i\hbar \frac{\partial}{\partial t} \Psi = \hat{H} \Psi$$

where $i\hbar \frac{\partial}{\partial t}$ is the *energy operator* (sometimes just written as $\hat{E}$), i is the imaginary unit, $\frac{\partial}{\partial t}$ is partial differentiation with respect to time, and $\hat{H}$ is the *Hamiltonian operator* associated with the system. The equation describes standing wave solutions, called *stationary states* in physics and *atomic orbitals* in chemistry.

Planck - Energy of a Quantum

$$E = h f$$

Momentum

$$p = mc$$

Speed of Light

$$c = f \lambda$$

$$i = \sqrt{-1}$$

deBroglie - Wavelength

$$\lambda = h / p$$

Wave number

$$\frac{2\pi}{\lambda} = \frac{p}{\hbar}$$

Einstein - Energy

$$E = mc^2$$
$$= pc$$

Kinetic Energy

$$E = \frac{1}{2} mc^2 = \frac{p^2}{2m}$$

☆

Dirac's Constant

$$\hbar = \frac{h}{2\pi}$$

Angular Velocity

$$\omega = 2\pi f = \frac{Et}{\hbar}$$

☆

☆

Wave function in Classical Mechanics

$$\psi = A\cos\left(\frac{2\pi r}{\lambda} - \omega t\right)$$

Euler's Equation

$$e^{i\theta} = \cos\theta + i\sin\theta$$

☆

☆

☆

☆

☆

☆

☆

☆ Schrödinger Wave Equation

$$i\hbar \frac{\partial}{\partial t} \psi(x,y,z,t) = \left(-\frac{\hbar^2}{2m}\nabla^2 + V(x,y,z)\right)\psi(x,y,z,t)$$

Energy Operator

Kinetic energy + potential energy   gives probability

Hamiltonian Operator

Simplified Form

$$\hat{E}\psi = \hat{H}\psi$$

∇^2 is the Laplacian operator $\dfrac{\partial^2}{\partial x^2} + \dfrac{\partial^2}{\partial y^2} + \dfrac{\partial^2}{\partial z^2}$

HIGHER DIMENSIONS
beyond the familiar knot

Many geometric forms in the two-dimensional plane can be extended to three dimensions by a process of analogy, circles, and spheres are a good example (*see page 66*). Although hard to visualize, we can easily continue this process, describing forms in four, five, or more dimensions (*e.g. below*). In general an n-dimensional space is given by the set of points with coordinates $(x_1, x_2, \ldots, x_n)$. Higher dimensional distances, angles, and other quantities can then be defined by analogy.

Einstein used four dimensions to model the physics of spacetime, and modern cosmologists use models of ten or more dimensions in string theory. It is important to remember, however, that an *n*-dimensional space can be defined and studied without any necessary interpretation in terms of physical space, time, or anything experiential.

Certain whole numbers *n* distinguish themselves in that the associated *n*-dimensional space has some unique geometric property. A simple example is the fact that three-dimensional space alone can support knots ('embeddings' of a circle). In any other number of dimensions any such embedding can always be 'unknotted'. Similarly, only four-dimensional space permits the mathematical curiosities known as *exotic differential structures*. But that's another story.

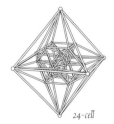

4-d simplex   *4-d orthoplex*   *tesseract (4-d cube)*   *24-cell*

BOOK III

Ingredients (makes one human): oxygen 61%, carbon 23%, hydrogen 10%, nitrogen 2.6%, calcium 1.4%, phosphorus 1.1%, potassium 0.2%, sulfur 0.2%, sodium 0.1%, chlorine 0.1%, plus magnesium, iron, fluorine, zinc, and other trace elements.

ESSENTIAL
ELEMENTS

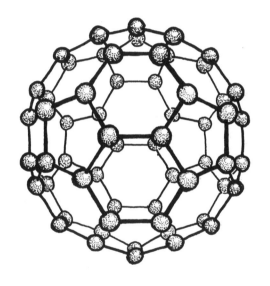

Matt Tweed

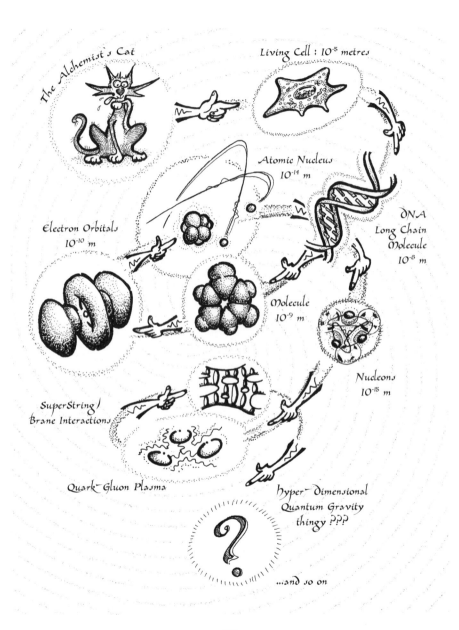

The Alchemist's Cat

Living Cell : 10^{-5} metres

Atomic Nucleus 10^{-14} m

DNA Long Chain Molecule 10^{-8} m

Electron Orbitals 10^{-10} m

Molecule 10^{-9} m

Nucleons 10^{-15} m

SuperString/ Brane Interactions

Quark-Gluon Plasma

hyper-dimensional Quantum Gravity thingy ???

?

...and so on

INTRODUCTION

Pretty much all that we see or touch in our seemingly solid existence is made from squintillions of tiny atoms, each being one of over a hundred unique types of element. Combined together in a myriad different ways, they form the fantastic mosaic that is the visible universe.

If we peer closely at an individual atom, the first astonishment is that it is mainly empty space. Fizzball electrons spin complex webs around a central nucleus, a miniscule point in the middle of a galaxy of whirling energy. Even here we are just scratching the surface—beyond are places where the rules become very strange indeed, where solidity has little meaning and matter comes in waves. Whole sub-atomic families appear, and particles interfere, tunnel, entangle and generally indulge in behaviours which defy common-sense, yet follow their own probabilistic laws. This is a place of high energies and fundamental forces that directly shapes our macroscopic experience.

Zooming in further still, we find that even this tiny kingdom may itself be the knottings of ever more ephemeral wisps on the very edges of our understanding, held together in symmetric patterns that span dimensions and perform a dance of deep mathematics.

All are players in this great game of life, acts of consciousness interlacing and resonating through the all that is.

Most of all I hope you, dear reader, will enjoy this brief journey into the wonderful world of matter. May we use this extraordinary knowledge with wisdom and understanding in the millennia ahead.

EARLY ALCHEMY
a wee bit of magick

The roots of chemistry stretch far back into the dim and distant past, to when our ancestors first prepared coloured earths for painting cave walls and themselves, learned the secrets of fire, and started experimenting with the arcane intricacies of cookery.

The ancient Egyptians knew of seven metals, as well as non-metals such as carbon and sulfur, all easily extracted from natural ores. The art of *Khemia,* supposedly revealed by angels, linked the metals to the seven known planets and assigned them unique qualities (*opposite top left*). Antique Indian treatises speak of the three *gunas,* fire, earth, and water. Chinese sages used two more, metal, and wood (*opposite top right*).

To the later Greek philosophers all things were made of earth, air, fire, or water (*opposite, lower left*). Naming them *elements,* Aristotle, in the 3rd century BC, added a fifth, *quintessence,* which formed the heavens. Another philosopher, Democritus, proposed that dividing matter over and over again would eventually leave an indivisible *atmos.* Scorned by Aristotle, the *atom* was then largely forgotten for centuries.

With the fall of the Greek empire, investigation of *Al-khemia* moved to Arabia. Books like Al-Razi's 10th century *The Secret of Secrets* and Jabir ibn-Hayyan's *The Sum of Perfection* told of an *elixir of life* that could grant immortality and transmute base metals into gold.

The quest spread to medieval Europe where, in the 13th and 14th centuries, alchemists like Albertus Magnus, Roger Bacon, and Nicholas Flamel hoped to find the all-powerful *Gloria Mundi* or *Philosopher's Stone.* Slowly, through experiment, trial, error, intuition, and the odd happy accident, they laid out the foundations of a extraordinary body of lore.

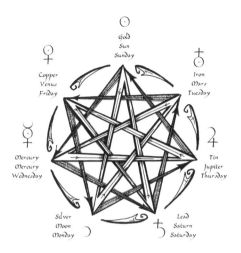

The Seven Metals of Antiquity
The Seven known Planets

Labels (clockwise from top):
Gold / Sun / Sunday
Iron / Mars / Tuesday
Tin / Jupiter / Thursday
Lead / Saturn / Saturday
Silver / Moon / Monday
Mercury / Mercury / Wednesday
Copper / Venus / Friday

Wu-hsing : the Five-Fold Chinese
Elemental System

Labels:
Wood / dragon / East / Spring / Green
Fire / Phoenix / South / Summer / Red
Earth / Snake / Center / Pivot / Ochre
Metal / Tiger / West / Autumn / White
Water / Tortoise / North / Winter / Black

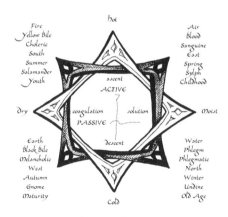

The Four Elements
and corresponding humours

Labels:
hot
Air / Blood / Sanguine / East / Spring / Sylph / Childhood
Moist
Water / Phlegm / Phlegmatic / North / Winter / Undine / Old Age
Cold
Earth / Black Bile / Melancholic / West / Autumn / Gnome / Maturity
dry
Fire / Yellow Bile / Choleric / South / Summer / Salamander / Youth

ascent
ACTIVE
coagulation   solution
PASSIVE
descent

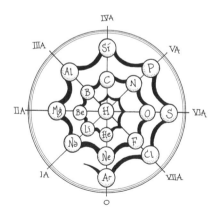

The Periodic Web of the
First Eighteen Elements

Labels: IVA, IIIA, VA, IIA, VIA, IA, VIIA, O
Elements: Si, Al, P, C, N, B, Mg, Be, O, S, H, Li, F, Na, He, Ne, Cl, Ar

127

THE AGE OF SCIENCE
alchemy transmutes into chemistry

By the eighteenth century scientists were largely freeing themselves of metaphysical concerns, and early experiments comparing weights and masses showed that many substances assumed to be elemental were in fact *molecules* or compounds of several parts.

In 1789 the first table of twenty-three elements was published by Antoine Lavoisier, soon followed by John Dalton's 1808 inspired forerunner to atomic theory (which was then ignored for fifty years).

As scientific techniques improved, new elements were discovered at a prodigious rate. Noticing how those with similar chemical properties fell into recurring patterns, Dmitri Mendeleev created his famed *Periodic Table of Elements* in 1869, successfully predicting the existence of scandium and germanium. The first hint of stuff smaller than the atom came in 1896, when, unwittingly leaving pitchblende (a uranium ore) on an unexposed photographic plate, Becquerel accidentally discovered radioactivity.

In the early twentieth century Ernest Rutherford's discovery of the surprisingly empty space around the atomic nucleus, the unveiling of the electron orbitals and Albert Einstein's theory that matter and energy were the same thing led Max Planck, Erwin Schrödinger, Niels Bohr, and others to the curiously wavy world of quantum mechanics. In 1932 the atom was split for the first time and for the rest of the century scientists explored the symmetries of the sub-atomic realms. Huge smashers hurled atoms together to synthesize new heavy elements, at other times breaking them apart to reveal whole families of exotic particles.

The universe was made of very strange things indeed.

95% Copper with 5% Tin heated to 1100 °C in charcoal furnace

Clay mould

Melt wax out

Wax Model

Molten bronze poured into mould

Break the mould

Polish and Finish

Bronze was one of the first alloys, often cast using the lost wax process

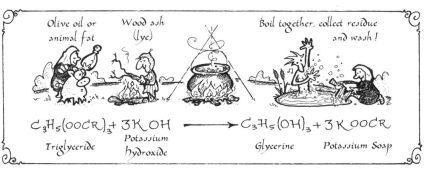

Olive oil or animal fat

Wood ash (lye)

Boil together, collect residue and wash !

$$C_3H_5(OOCR)_3 + 3K\,OH \longrightarrow C_3H_5(OH)_3 + 3K\,OOCR$$

Triglyceride

Potassium hydroxide

Glycerine

Potassium Soap

Soaps were perhaps discovered from fat falling into the dampened ashes of a fire

Humphrey Davy
Group I & II Elements

Dimitry Mendeleev
Periodic Table

Ramsay and Travers
Noble Gases

Marie and Pierre Curie
Radium & Polonium

Berkeley and Dubna scientists
Transuranium Elements

Some of the many involved in the discovery of the elements

INSIDE THE ATOM
protons, neutrons, and electrons

Atoms consist of a small central *nucleus* orbited by one or more whirling *electrons*. Two visualizations of an atom are shown opposite. The nucleus, a mere hundred billionth of a millimeter across, contains two similarly sized particles, *protons* and *neutrons*. Each proton has a single positive electric charge and a corresponding negatively charged electron. It is the proton count which gives an element its name and *atomic number*, or position in the periodic table (*see pages 390–391*). Neturons have no charge.

Although the number of protons and electrons are fixed in every element, the number of neutrons can vary, giving *isotopes*, which react the same chemically yet can behave quite differently at a nuclear level.

Electrons weigh almost two thousand times less than protons and neutrons. Repelling other negative electrons they are attracted to the oppositely charged positive protons but ignore the chargeless neutrons. Balancing all these forces, electrons team up in pairs and whizz about the nucleus in *orbitals* grouped into orbital *sets*, three dimensional patterns that get increasingly complex in larger atoms. Orbitals fill up in a specific order (*lower opposite*).

Amazingly, atoms are almost entirely empty space. An electron orbiting a nucleus may be visualized as a cat swinging a bumblebee on the end of a half-mile long piece of elastic.

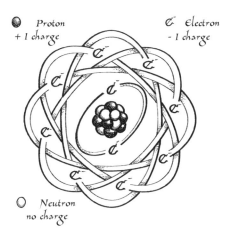

Proton
+ 1 charge

Electron
- 1 charge

Neutron
no charge

A classical planetary picture of a neon atom :
a central nucleus of 10 neutrons and 10
protons surrounded by a whirl of 10 electrons,
two in an inner orbit, the remainder in an outer.

A quantum mechanical view of the same atom :
here each lobe represents the probability of finding
an electron in a particular place as given by the
Schrödinger wave equation.

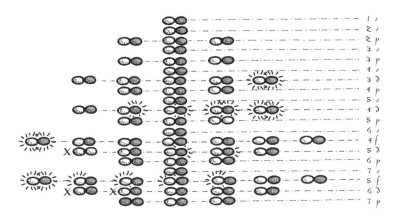

1 s
2 s
2 p
3 s
3 p
4 s
3 d
4 p
5 s
4 d
5 p
6 s
4 f
5 d
6 p
7 s
5 f
6 d
7 p

Electron orbitals build in sequence from the innermost 1s. Each row half fills with electrons (white blobs)
before completing as oppositely spinning pairs (black blobs). Glows around a blob indicate that one or both
electrons skip to or from other orbitals, breaking the pattern. Gold, silver and copper are amongst those that
share this quirk. X marks the d- orbitals that try to start filling before the row above gets going.

Periodic Tables

elemental ordering

Every element has its own place in the periodic table, and there are several versions of the table that emphasize different features.

Professor Benfey's spiral (*below*) develops by atomic number (the count of protons in each element) and shows *groups* with the same pattern of outer electrons (and hence corresponding properties) radiating like spokes from a hydrogen hub. As the different orbitals fill, *blocks* of related elements form outcrops.

In a contrasting scheme, Dr Stowe's table (*opposite top*) displays the physical ordering of the intricate orbital sets of electron *shells*, with the innermost at the top, using each element's unique set of *quantum numbers*.

A modern version of Mendeleyev's original table (*lower opposite*) puts groups in vertical columns with horizontal *periods* of orbital sets. Elements are arranged by atomic number, reading left to right, top to bottom.

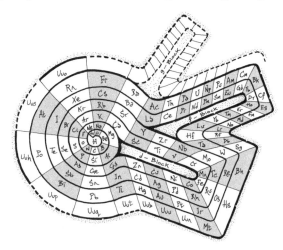

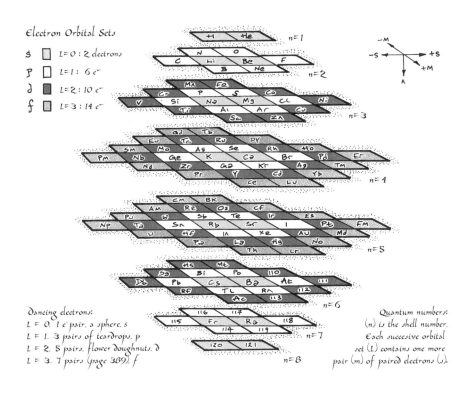

Electron Orbital Sets

s ▢ L=0 : 2 electrons
p ▢ L=1 : 6 e⁻
d ▢ L=2 : 10 e⁻
f ▢ L=3 : 14 e⁻

Dancing electrons:
L = 0. 1 e⁻ pair, a sphere, s
L = 1. 3 pairs of teardrops, p
L = 2. 5 pairs, flower doughnuts, d
L = 3. 7 pairs (page 389), f

Quantum numbers:
(n) is the shell number.
Each successive orbital
set (L) contains one more
pair (m) of paired electrons (s).

Above: The Atomic Shells, with the shaded orbital sets of which they are comprised.
Below: The modern periodic table, which is read left to right across the open spaces and shows
increasing order of atomic number (see pages 390–391).

133

A BURNING QUESTION
chemical conflagrations

Most things around us are *compounds*, or combinations of various elements. To bond, atoms rearrange their outer, highest energy *valence* electrons, keeping their full interiors safely out of the way.

Opposite top is an *exothermic* reaction, which produces heat. After initially shaking apart the gas molecules with the heat of a flame, new water bonds quickly form which lock up less energy than the original gas bonds. The released energy keeps the reaction going and the gases explode, frantically shuffling electrons between each other. Lower opposite we see an *endothermic* reaction that takes place when plants *photosynthesize*. Here the sums go the other way round and heat needs to be absorbed, in this case from sunlight. The products therefore have more energy than the reactants, glucose storing the energy. When the reaction moves in the opposite direction, it is known as *respiration*.

Matter itself exists in several different states or *phases (below)*. *Solids* pack atoms closely in rigid arrangements. Heating vibrates the atoms, shaking the structures apart to form *liquids* that can flow and change shape. Heating further weakens even these loose bonds and the atoms scatter at high speed in all directions as *gases*. At yet higher temperatures some of the electrons are knocked off the atoms to create an electrically charged, ionized *plasma*, like that found in the superhot corona of the sun.

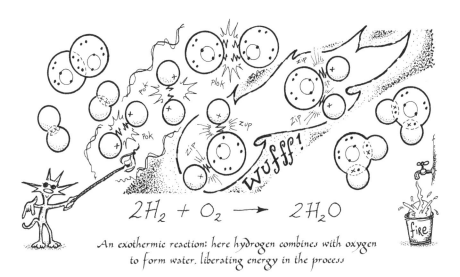

$$2H_2 + O_2 \longrightarrow 2H_2O$$

An exothermic reaction: here hydrogen combines with oxygen to form water, liberating energy in the process

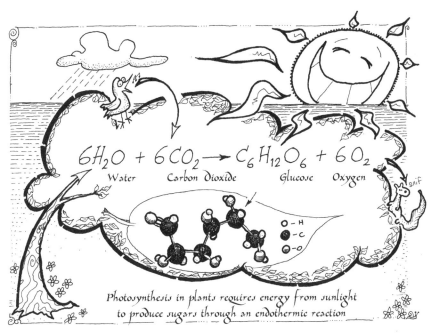

$$6H_2O + 6CO_2 \longrightarrow C_6H_{12}O_6 + 6O_2$$

Water   Carbon Dioxide   Glucose   Oxygen

○ – H
● – C
◐ – O

Photosynthesis in plants requires energy from sunlight to produce sugars through an endothermic reaction

BONDING

atomic stickiness

Molecules are formed as atoms' outer electrons share dances. Losing or gaining electrons causes atoms to become positively or negatively electrically charged *ions*. Most elements are either metals, which are *electropositive*, losing electrons to form *cations*, or non-metals, which are *electronegative*, grabbing electrons to form *anions*.

An *ionic bond* occurs when a positive cation lends electrons to a negative anion to give them both full stable outer orbitals like the nearest noble gas (*opposite top left*). Though tough and brittle with high melting points, many ionic compounds dissolve in water.

Non-metals combine using *covalent bonds*, which shuffle and share outer electrons into pairs, again filling up any empty orbitals (*opposite top right*). The attraction felt by electrons for nuclei, outweighing their mutual repulsions, holds the resulting molecule together.

In *metallic bonds* electrons float away from their nuclei, *dissociating* into a 'sea' around a lattice of positive ions (*opposite*). The conductivity and shininess of metals is the direct result of these mobile electrons, and their strength and high melting points result from the lovestruck relationship between the ions and their mates.

Hydrogen attached to a non-metal pushes against unbonded *lone pair* electrons creating a very slight charge difference across the molecule. If another electronegative atom is nearby, a weak *hydrogen bond*, vital in water and DNA, appears between them (*opposite lower left*).

With asymmetrical motions of electrons causing instantaneous small *Van der Waals forces* between atoms, and overlapping orbitals smearing π-bonds (*opposite lower right*), atomic glues come in many forms.

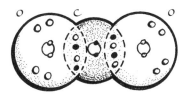

Lithium Flouride Ionically Bonding

Covalent Bonding in Carbon Dioxide

Pseudo-Electron Density Map of Crystalline Lithium Flouride
(Nuclei centres are 200.9 pico-metres apart)

The Noble Ship of Current Sails the Metallically Bonded Dissociated Sea of Electrons

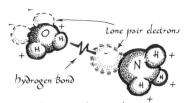

Lone pair electrons

hydrogen Bond

hydrogen Bonding between
Water (H₂O) and
Ammonia (NH₃)

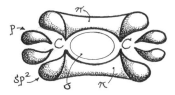

π-Bonds formed from overlapping orbitals
alongside a covalent σ-bond create the
Carbon double bond in ethylene

137

CRYSTALS
building bigger

Crystals are simple repeated patterns of unit cells. Like apples on a market stall, atoms and molecules can grow into large three-dimensional solids, the pieces positioned to give the best balance between attractive and repulsive forces. Ordered in their zillions, these tiny blocks build into many of the solid substances of our world, bridging the vast difference of scale between molecules, minerals, and mountains.

There are seven crystalline systems based on tessellating geometries which, when combined with the four basic unit cell types, give the fourteen *Bravais Lattices* (*opposite*). Variations in temperature or pressure may change one crystal structure into another more comfortable and efficient arrangement. Sulfur, for example, transforms from an orthorhombic lattice to monoclinic at 96°C, quickly reverting back on cooling.

More complex semi-regular or aperiodic crystalline systems also occur, in living things (*e.g. below left*) and in quasicrystals, rapidly cooled metal alloys (*e.g. the five-fold Al-Mn system below right*).

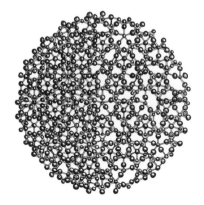 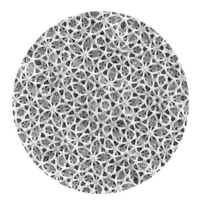

Simple Cubic

Body-Centered Cubic

Hexagonal System

Face-Centered Cubic

Simple Monoclinic

Base-Centered Monoclinic

Rhombohedral System

Body-Centered Tetragonal

Simple Tetragonal

Body-Centered Orthorhombic

Simple Orthorhombic

Face-Centered Orthorhombic

Triclinic System

Base-Centered Orthorhombic

HYDROGEN AND HELIUM
the first two elements

Hydrogen makes up three quarters of all known matter in the universe and is a large part of most stars. The first element, and the simplest atom, it consists of one proton orbited by one electron.

Hydrogen gas is *diatomic*, which means it is happiest when two atoms covalently bond to form one molecule, H_2. Highly explosive in air, it burns rapidly with oxygen to create water. Under immense pressures and temperatures (everyday conditions in the cores of giant planets like Jupiter and Saturn) hydrogen becomes metallic (*see page 354*).

The second element in the periodic table is *helium*. It has two protons, two electrons, two neutrons (99.99% of the time), and is the second most abundant element in the material universe, almost a quarter of it. With two electrons completely filling the *1s*-orbital, helium is happy to stay independent and rarely reacts with other elements. It is the first of the *noble* (or *inert*) gases, each of which have full outer electron orbitals. Surprisingly, helium was unknown on Earth until 1870 when it was discovered through *spectrographic analysis* of sunlight, a fingerprinting technique for elements (*lower opposite*). Lighter than air, though twice as heavy as hydrogen, the specks of helium formed here quickly float off into outer space. It is a much safer gas than hydrogen in balloons, and when inhaled produces a squeaky voice.

Beyond its common form, hydrogen has two isotopes, *deuterium*, with one neutron, and *tritium*, with two. Tritium is a rare and unstable beast, decaying into the light helium isotope *helium-3* by changing a neutron into a proton via radioactive beta decay (*see page 160*).

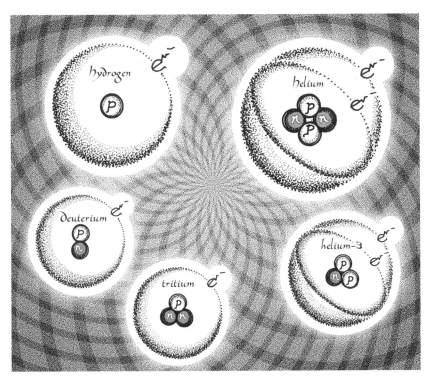

hydrogen and helium: with two isotopes of hydrogen and one of helium
showing the number of protons (p), neutrons (n) and electrons (e⁻)

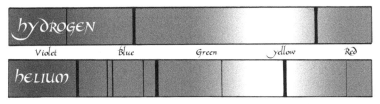

Each element absorbs light in a unique way producing dark bands at
specific places across the electromagnetic spectrum: in this way even
very distant stars can be analyzed to discover their chemistry.

ALKALI & ALKALINE-EARTH METALS
the violent world of the s-block

The first real group of the periodic table is known as the *alkali metals*, IA *(leftmost column, below)*. Soft and silvery-white, they all have a single outer *s*-orbital electron which they enthusiastically lose to form singly charged +1 ions, making them very electropositive.

Lithium, the first member of the group, and the third element, is the lightest metal and floats on water. *Sodium*, immediately below, floats and fizzes as it oxidizes, regularly bonding with chlorine to make common salt (NaCl). Next down, *potassium* is the second lightest metal, oxidizing rapidly in air and bursting into flames when wet. Lower still, both *Cæsium*, the most electropositive element, and *rubidium* explode on contact with air. *Francium*, the final member of this vigorous family, is radioactive.

Moving across one column, we meet group IIA, the rare earth metals, *beryllium, magnesium, calcium, strontium, barium,* and *radium.* Marginally less electropositive, they gladly form double-charge +2 ions, losing both their outer electrons. They are denser than their group I neighbours, with higher melting and boiling points.

A wire dipped in compounds of these elements will produce characteristic colours when held in a flame. Excited electrons jump between orbitals, losing their energy as little packets of light, *photons,* on the way back down to their normal state *(lower opposite)*.

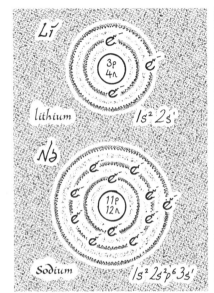

Li

lithium

3p
4n

$1s^2 2s^1$

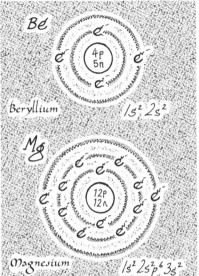

Be

Beryllium

4p
5n

$1s^2 2s^2$

Na

Sodium

11p
12n

$1s^2 2s^2 p^6 3s^1$

Mg

Magnesium

12p
12n

$1s^2 2s^2 p^6 3s^2$

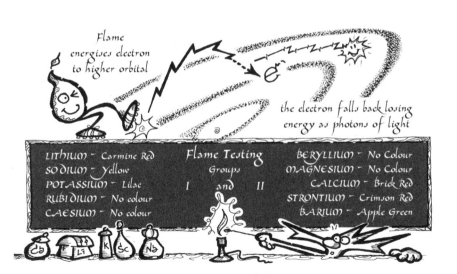

Flame energises electron to higher orbital

the electron falls back losing energy as photons of light

	Flame Testing	
LITHIUM – Carmine Red		BERYLLIUM – No Colour
SODIUM – Yellow	Groups	MAGNESIUM – No Colour
POTASSIUM – Lilac	I and II	CALCIUM – Brick Red
RUBIDIUM – No colour		STRONTIUM – Crimson Red
CAESIUM – No colour		BARIUM – Apple Green

THE P-BLOCK

metals, metalloids, and non-metals

Elements five to ten are the first members of the *p-block*. One to six electrons inhabit three new double-teardrop shaped *p*-orbitals arranged at right angles around the nucleus (*see page 162*). At room temperature and pressure they appear as solids (carbon and aluminium), liquids (bromine), and gases (nitrogen and chlorine), depending on the balance of their interatomic and intermolecular forces. The left-hand side of the *p*-block mostly shines with solid metals. Ductile, malleable, and conductive because of their footloose outer electrons, most metals can be stretched into wires, squished into sheets, or combined into alloys.

Crossing the block from left to right, we move from metals to non-metals. These tend to be dull brittle solids, liquids, or gases that are poor conductors of heat and electricity. In this small corner are found many of the players the game of life, such as carbon, oxygen, and nitrogen, whose compounds form the backbone of living things and organic chemistry (*opposite top, and page 152*).

In between metals and non-metals lie the *metalloids*, a diagonal streak of ambiguous elements with aspects of both. Among these are the semiconductors, boron, silicon, germanium, and arsenic, which form the minds of our computers and electronic gizmos due to their outer valence electrons' ability to jump about inside their nearly full shells.

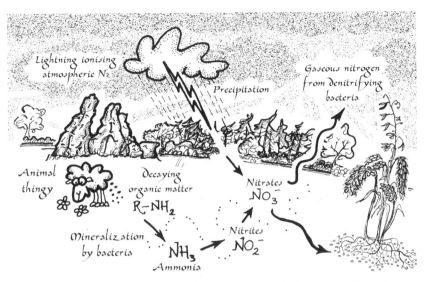

Lightning ionising atmospheric N₂

Precipitation

Gaseous nitrogen from denitrifying bacteria

Animal thingy

Decaying organic matter
$R-NH_2$

Nitrates
NO_3^-

Mineralization by bacteria

Nitrites
NO_2^-

NH_3
Ammonia

To be useful to plants the strong N₂ triple bond has to be broken by nitrogen fixing bacteria

5 6	6 6	7 7	8 8	9 10
B +3	**C** -4+2+4	**N** -3+1+5	**O** -2	**F** -1
13 14	14 14	15 16	16 16	17 18
Al +3	**Si** -4+2+4	**P** -3+1+5	**S** -2+4+6	**Cl** -1
31 38	32 42	33 42	34 46	35 44
Ga +3	**Ge** +2+4	**As** +3+3+5	**Se** -2+4+6	**Br** -1+1+5
49 66	50 70	51 70	52 78	53 74
In +3	**Sn** +2+4	**Sb** -3+3+5	**Te** -2+4+6	**I** -1+1+5+7
81 124	82 126	83 126	84 125	85 125
Tl +1+3	**Pb** +2+4	**Bi** +3+5	**Po** -2+4	**At** ?
113	Y114	115	116	117
Uut	**Uuq**	**Uup**	**Uuh**	**Uus**

Metalloids run diagonally across the p-block marking the changeover between metals and non-metals.

Boron isn't boring
I can tell you that
This bug isn't resting
It's quite dead out flat
All ready for the cooking in
A flameproof pyrex dish
of
Borosilicates !!
The fifth atom's
special wish.

BORON IS NOT BORING

e^- e^- e^-
e^-

CARBON AND SILICON
organic and virtual thinking materials

Twenty-three percent of you is carbon. The sixth element underpins *organic* chemistry, the fabric of life, from DNA and proteins in our cells to once living stuff, plastics, and fossil fuels. Coming in a dazzling array of molecules, carbon is neither electropositive or negative. A non-metal, it combines with many other elements and also extensively with itself, creating long chains and rings (*see page 388*). Multiple π-bonds smear electrons between atoms to give double and triple bonds.

Carbon arranges itself into several different *allotropes*. In diamond crystals every atom bonds to four others in a hard tetrahedral grid (*opposite top right*) whereas in graphite, a soft crystalline solid found in charcoal and pencils, flat planes of carbon rings slide easily over each other (*opposite top left*). Each atom here joins to three others, the π-bonds enabling it to conduct electricity. Other allotropes include spherical *buckminsterfullerenes*, intriguing *nanotubes*, and *graphene*, all of which have amazing structural and conductive properties.

Directly beneath carbon in the periodic table is *silicon*, a metalloid semi-conductor. Carbon life is mirrored by silicon logic in the buzzing microchip mazes of purified silicon crystals, doped with elements like gallium or arsenic to alter their electronic properties.

Stable silicon compounds cover much of the earth as rocks and minerals, such as quartz-rich sand and fine-grained aluminium silicate clays. The Earth's crust consists of 60% aluminium silicate *feldspars* ($KAl Si_3O_8$-$NaAlSi_3O_8$-$CaAl_2Si_2O_8$), followed by quartz (SiO_2), then olivine ($(Mg,Fe)_2SiO_4$ (*see appendix page 402*). Clays have unusual life-mimicking chemistries and perhaps contributed to biological evolution.

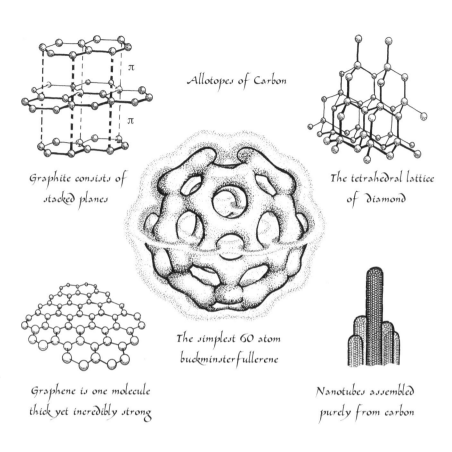

Allotopes of Carbon

Graphite consists of
stacked planes

The tetrahedral lattice
of diamond

The simplest 60 atom
buckminster fullerene

Graphene is one molecule
thick yet incredibly strong

Nanotubes assembled
purely from carbon

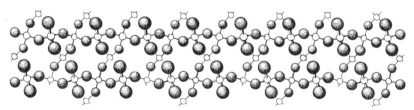

In silicon dioxide (SiO_2) each silicon atom is tetrahedrally bonded to four
oxygen atoms to create quartz crystals and the sand on our beaches.

OXYGEN AND SULFUR
the over and underworlds of group VI A

A fifth of the air we breathe is *oxygen*. After hydrogen and helium, it is the third most abundant element in the universe. Highly reactive, oxygen exerts a strong pull on other atoms to gain the two electrons needed to fill its outer orbitals. Redox (reduction-oxidation) reactions play a fundamental role on our planet and are vital to life, involved in everything from metabolizing food to sending nerve impulses.

On Earth, free oxygen forms an odourless diatomic gas (O_2). High in the atmosphere, naturally occurring triatomic *ozone* (O_3) protects us from harmful ultraviolet radiation, yet lower down its powerful oxidizing effects can damage living organisms. The ten most common compounds in the Earth's crust are *oxides*; just under half is sand (silicon dioxide SiO_2), a third is magnesium oxide (MgO), and much of the rest is rock salt, iron(II) oxide (FeO). Water (H_2O) is another essential oxygen compound, covering 71% of our world's surface.

Below oxygen in the periodic table, so mirroring its chemistry in many ways, is the smelly underworld of *sulfur*. Usually a brittle, pale yellow solid, it has a profusion of multi-atom ring and chain allotropes, burning in air to create sulfur dioxide (SO_2). Combining with water in clouds, it becomes tree-scolding sulfurous acid rain. Sulfur is less electronegative than oxygen, and hydrogen sulfide (H_2S) acts differently than water, hydrogen bonding having little influence. To us a fiendishly toxic gas with a rotten egg smell, colonies of creatures nevertheless live in the dark on energy metabolized by sulfur-breathing bacteria beside deep ocean volcanic vents bubbling hydrogen sulfide. Indeed, life on Earth may have started down in hot wet places like these.

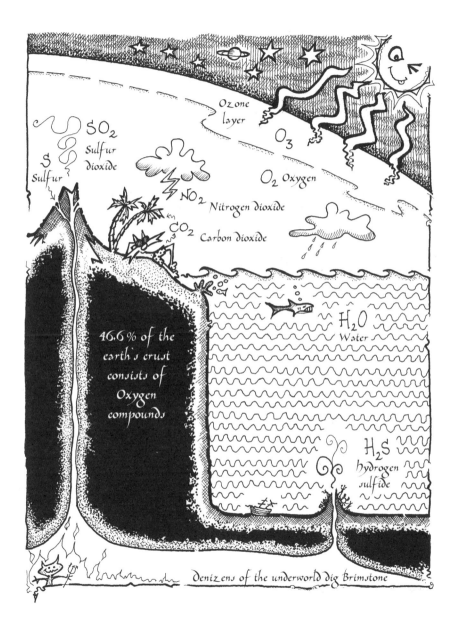

Ozone layer

O_3

SO_2
Sulfur dioxide

S
Sulfur

O_2 Oxygen

NO_2 Nitrogen dioxide

CO_2 Carbon dioxide

46.6% of the earth's crust consists of Oxygen compounds

H_2O
Water

H_2S
hydrogen sulfide

Denizens of the underworld dig Brimstone

WATER AND ACIDS
making a splash

Water is the most common molecule in the universe. One oxygen and two hydrogen atoms, H_2O is two thirds of all of us.

The water molecule is *polarized*. The pull from the oxygen atom gives the hydrogen atoms a slight positive charge (*opposite top left*) resulting in extensive networks of hydrogen-bonded molecules creating sixfold snow crystals (*opposite*), surface tension, and the fluctuating crystal lattice we drink (*below*). Water's attraction allows it to unsettle and dissolve even its own molecules into ions, one water molecule donating a proton (H^+) to another, forming a solution of hydronium (H_3O^+) and hydroxide (OH^-) ions.

Acids are compounds that actively donate protons when in solution, attacking metals to liberate hydrogen gas. Gladly accepting the protons, *bases* are compounds which are soapy and bitter, combining with an acid to form a *salt* and water. An everyday example is hydrochloric acid and sodium hydroxide combining into common salt and water: $HCl + NaOH = NaCl + H_2O$ Certain metal oxides, hydroxides, amines, and the group I and II alkalis are particularly caustic in this respect. Other *Lewis* acids and bases respectively accept or donate an electron pair subject to the solvents used, and not all acids need water to do their corrosive work.

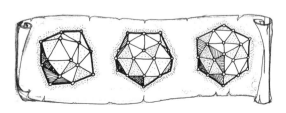

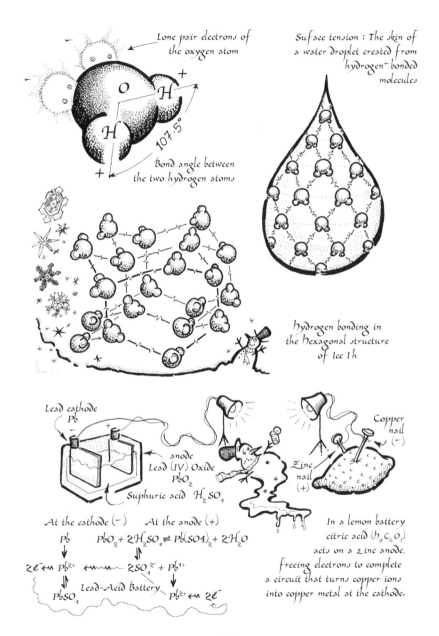

Lone pair electrons of the oxygen atom

Suface tension : The skin of a water droplet created from hydrogen-bonded molecules

O

H   H

$107.5°$

+   +

Bond angle between the two hydrogen atoms

hydrogen bonding in the hexagonal structure of Ice Ih

Lead cathode Pb

+

anode Lead (IV) Oxide PbO_2

Suphuric acid H_2SO_4

Copper nail (-)

Zinc nail (+)

At the cathode (-)   At the anode (+)

Pb $PbO_2 + 2H_2SO_4 \rightleftharpoons Pb(SO4)_2 + 2H_2O$

$2e^- \leftarrow Pb^{2+} \leftarrow \sim \sim \cdots 2SO_4^{2-} + Pb^{4+}$

$PbSO_4$ Lead-Acid Battery $Pb^{2+} \leftarrow 2e^-$

In a lemon battery citric acid ($h_8C_6O_7$) acts on a zinc anode, freeing electrons to complete a circuit that turns copper ions into copper metal at the cathode.

ORGANIC CHEMISTRY
biocosmic molecules

Carbon creates literally millions of compounds and is vital to life as we know it. *Hydrocarbons* are made only of carbon and hydrogen and form the basis of the oils that power industry. Crude oil, a mix of hydrocarbons, is separated out into fractions through a heated column (*opposite top right*), in which the lighter oils with fewer carbon atoms rise the highest. Organic chemistry deals with combinations of carbon and hydrogen, often with oxygen and nitrogen, sulfur, phosphorus, and others.

Easily linking into lattices and sharing bonds gives carbon the ability to *polymerize*, joining simple *monomer* molecules into long chains of repeating units (*opposite top left*). Polymers are the foundations of the fantastic plastics that are used everywhere from satellites to false teeth.

The building block of organic chemistry are the *functional groups* which define how compounds behave (*see page 388*). Methanol (CH_3OH), the first alcohol (-OH group), is poisonous whilst ethanol (C_2H_5OH), the second, can liven up an evening. Change the alcohol into an aldehyde (-CHO), however, and it can ruin the next morning (*lower opposite*).

Carbon molecules occur in vast quantities in space, drifting in giant molecular clouds that form the nurseries of new stars. Natural reactions turn simple molecules such as carbon monoxide (CO), water (H_2O), ammonia (NH_3), methane (CH_4), cyanic acid (HOCN), and formaldehyde (H_2CO) into more complex ones like acetone (($CH_3)_2CO$), ethyl alcohol (CH_3CH_2OH), and cyanodecapentayne ($HC_{10}CN$). Even some fullerenes (e.g. C_{70}) float free. Hydrocarbons are abundant on other planets, raining down to form lakes on Saturn's moon Titan. Interstellar laboratories often create the essential elements of life, so perhaps we are not alone.

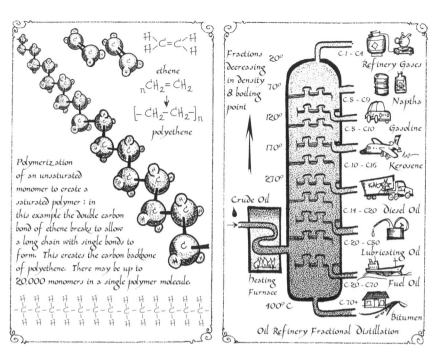

H₂C=CH₂

ethene

$$n\,CH_2 = CH_2$$
$$\downarrow$$
$$[-CH_2-CH_2-]_n$$

polyethene

Polymerization of an unsaturated monomer to create a saturated polymer: in this example the double carbon bond of ethene breaks to allow a long chain with single bonds to form. This creates the carbon backbone of polyethene. There may be up to 20,000 monomers in a single polymer molecule.

Fractions decreasing in density & boiling point

20°	C1 - C4	Refinery Gases
70°	C5 - C9	Naptha
120°	C5 - C10	Gasoline
170°	C10 - C16	Kerosene
270°	C14 - C20	Diesel Oil
	C20 - C50	Lubricating Oil
	C20 - C70	Fuel Oil
400°C	C70+	Bitumen

Crude Oil

heating Furnace

Oil Refinery Fractional Distillation

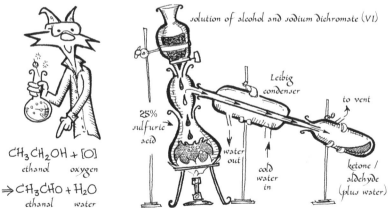

solution of alcohol and sodium dichromate (VI)

Leibig condenser

to vent

25% sulfuric acid

water out

cold water in

ketone / aldehyde (plus water)

$$CH_3CH_2OH + [O]$$
ethanol   oxygen

$$\Rightarrow CH_3CHO + H_2O$$
ethanal   water

Recipe for a hangover: preparing an aldehyde/ketone from alcohol by heating under reflux. The dichromate (VI) is reduced by the alcohol to a chromium (III) ion, whilst the alcohol is oxidized to an aldehyde or ketone.

HALOGENS & THE NOBLE GASES
ups and downs at period's end

The universe's most vigorous and inert elements are found in the final two columns of the periodic table. The members of group VIIA, the *halogens*, are just one electron short of a full shell, and aggressively form compounds to complete it. All elements, except helium, neon, and argon, bond with a halogen to form a *halide*.

The ninth element, and easily the most electronegative, is *fluorine*, a pale green-yellow diatomic gas which combines fanatically with almost anything, attacking compounds to form *fluorides*. The rest of the group are also intensely reactive, particularly that rascal chlorine, which is why it is so good at killing bugs in bleaches (*see opposite*).

With one more proton and one more electron added, we finally meet the quiet, solipsistic group VIIIA. With all the slots of their electron orbitals full, they are closed to business and on the whole content not to react with anything. That said, *xenon* does form (with effort) a few compounds with feisty fluorine and its neighbour, oxygen, and a few helium and krypton compounds also exist, so the former name of this group, the *inert gases*, was changed in the 1960s to the slightly less lazy *noble gases*. When you next see glowing neon signs, picture those full orbitals frantically buzzing with jumping electrons.

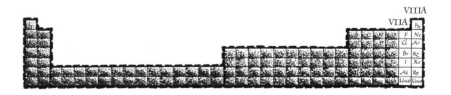

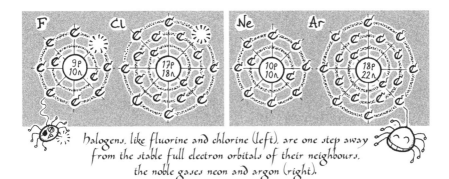

halogens, like fluorine and chlorine (left), are one step away
from the stable full electron orbitals of their neighbours,
the noble gases neon and argon (right).

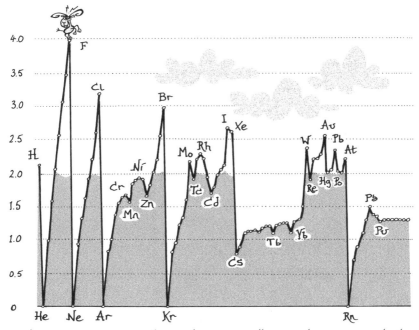

Electronegativity measures how easily an atom will attract electrons in a molecule:
the very reactive halogens occupy the highest peaks whilst the noble gases
(with the notable exception of xenon) sleep quietly in the deepest valleys.

THE TRANSITION METALS
gold, silver, copper, and iron in the d-block

The next zone we encounter on our travels across the periodic kingdom is a series of metals starting at scandium where the first of ten electrons begins filling the set of *3d* orbitals *inside* the *4s* (*see page 133*). Most members of this series heartily lose one or more electrons to form a bewildering array of brightly coloured compounds.

The *transition metals* are hard and strong, their similar structures allowing them to be mixed into useful *alloys*. Copper and zinc combine into brass, and mercury, the only metal liquid at room temperature, forms alloys called *amalgams*. With a dash of carbon, iron creates steel, becoming even harder with an added splash of vanadium, molybdenum, or chromium. Iron's magnetic attraction is due to the unbalanced magnetic moments of unpaired electrons in its outer *d*-orbital. Several near neighbours, notably nickel, cobalt, and manganese, also exhibit varying degrees of paramagnetism.

Titanium has a reputation for both strength and corrosive resistance, and is thus ideally suited for flying machines and rocket ships.

The enduring popularity of shiny gold, silver, and copper is in part due to their dependable stability. Excellent conductors of electrons and heat, they have many applications in electronics and optics, also looking very pretty in rings, crowns, coins, and other baubles.

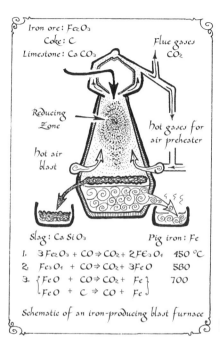

Iron ore: Fe_2O_3
Coke: C
Limestone: $CaCO_3$
Flue gases CO_2

Reducing Zone

hot gases for air preheater

hot air blast

Slag: $CaSiO_3$　　Pig iron: Fe

1. $3Fe_2O_3 + CO \Rightarrow CO_2 + 2Fe_3O_4$　450 °C
2. $Fe_3O_4 + CO \Rightarrow CO_2 + 3FeO$　580
3. $\begin{cases} FeO + CO \Rightarrow CO_2 + Fe \\ FeO + C \Rightarrow CO + Fe \end{cases}$　700

Schematic of an iron-producing blast furnace

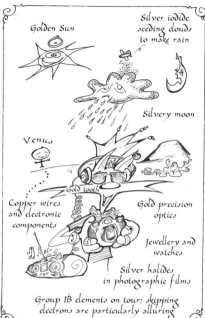

Golden Sun

Silver iodide seeding clouds to make rain

Silvery moon

Venus

Gold tooth

Copper wires and electronic components

Gold precision optics

Jewellery and watches

Silver halides in photographic films

Group IB elements on tour: skipping electrons are particularly alluring

Z	electrons (d-s)	Symbol	Oxidation states
21	1-2	Sc	+3
22	2-2	Ti	+2 +3 +4
23	3-2	V	+2 +3 +4 +5
24	5-1	Cr	+2 +3 +6
25	5-2	Mn	+2 +3 +4 +7
26	6-2	Fe	+2 +3
27	7-2	Co	+2 +3
28	8-2	Ni	+2 +3
29	10-1	Cu	+1 +2
30	10-2	Zn	+2
39	1-2	Y	+3
40	2-2	Zr	+4
41	4-1	Nb	+3 +5
42	5-1	Mo	+6
43	5-2	Tc	+4 +6 +7
44	7-1	Ru	+3
45	8-1	Rh	+3
46	10-0	Pd	+2 +4
47	10-1	Ag	+1
48	10-2	Cd	+2
57	1-2	La	+3
72	2-2	Hf	+4
73	3-2	Ta	+5
74	4-2	W	+6
75	5-2	Re	+4 +6 +7
76	6-2	Os	+3 +4
77	7-2	Ir	+3 +4
78	9-1	Pt	+2 +4
79	10-1	Au	+1 +3
80	10-2	Hg	+1 +2
89	1-2	Ac	+3
104	?	Rf	+4
105	?	Db	?
106	?	Sg	?
107	?	Bh	?
108	?	Hs	?
109	?	Mt	?
110	?	Uun	?
111	?	Uuu	?
112	?	Uub	?

The d-block transition metals, showing for each element the number of protons (top left), the number of electrons in the outer d- and s-orbitals (top right), and oxidation states (below). Electrons skip from outer s- to d-orbitals when the latter are half-full or full.

THE F-BLOCK AND SUPERHEAVIES
enormous atoms and islands of stability

At lanthanum, element fifty-seven, a *5d* orbital starts to fill before something strange happens; the next electron drops into a previously hidden *4f* orbital inside the full *6s*, *5s*, and *5p* orbital sets, taking the electron from the *5d* with it. The *5d* orbitals wait patiently until the *4f*s are full, apart from one halfway hiccup at gadolinium, where an electron briefly flickers up to the *5d*.

Quietly spreading from lanthanum to lutetium the *lanthanides*, or *rare-earth* metals, fill a fourteen place set of *4f* orbitals. Overshadowed by their *5s* and *5p* sets, only subtle chemical differences are found in this series.

Below the lanthanides, the radioactive *actinides* play much the same trick, as two electrons begin a *6d* orbital only to quit the job and turn within to fill a *5f* orbital instead. Uranium is the last natural element; artificially made atoms now fill a seventh of the periodic table.

Beyond the *f*-block at rutherfordium, a fourth transition series starts filling a *6d* orbital. These superheavy elements tend to be highly radioactive and unstable due to uhappy ratios of protons to neutrons (*see map opposite*). Elements with up to 118 protons have been fleetingly created in particle accelerators. Around elements *114p* 'eka-lead' or *184n* there may possibly be rare islands of stability where a few isotopes with balanced nuclei have significant lifetimes of minutes rather than seconds.

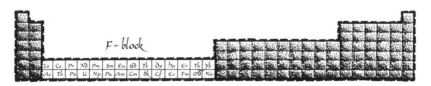

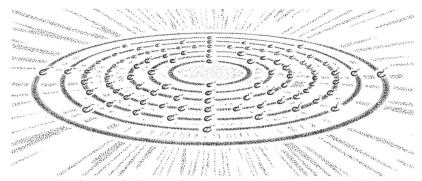

The electrons of Plutonium (not the order they fill in - see page 133)
$1s^2 \ 2s^2 \ 2p^6 \ 3s^2 \ 3p^6 \ 3d^{10} \ 4s^2 \ 4p^6 \ 4d^{10} \ 4f^{14} \ 5s^2 \ 5p^6 \ 5d^{10} \ 5f^6 \ 6s^2 \ 6p^6 \ 7s^2$

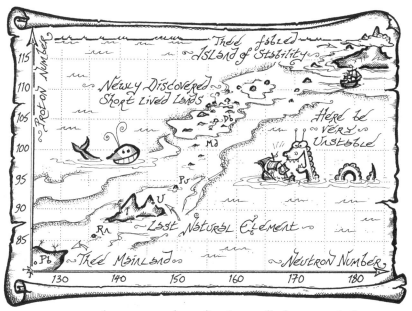

A Mappe to show ye wayye to lands afar whence stable elementes might forme
wythe a balancing of neutron and protonne numbers

RADIOACTIVITY
nuclear fizzicks

Held together by immensely strong forces, an atomic nucleus contains huge amounts of energy. Unstable nuclei rebalance by spitting out radioactive emissions as protons and neutrons join (*fusion*), or split off (*fission*). An isotope's radioactivity halves over its *half-life*; the less stable it is the faster it *decays*. Uranium-238 has a half-life of 4.5 billion years, yet with ten neutrons less, uranium-228 halves every fifth of a second!

All living things are slightly radioactive, constantly absorbing carbon-14 and tritium generated by cosmic rays. At death, we stop gathering these isotopes, and archæologists use the 5,730-year half-life of carbon-14 to date historically interesting goo.

Beyond bismuth, all elements have *radioisotopes* that undergo α-*(alpha)* decay, the nucleus expelling an α-*particle* (a helium nucleus, *see page 141*). Thin clothing should prevent α-particles from ionizing the unwary. Excess neutrons in a nucleus cause β-*(beta)* decay, where a neutron converts into a proton, releasing a speeding electron (a β-*particle*); protective apparel or 2 mm of aluminium halts these beasties. Often found alongside α- or β-decays, γ-*(gamma) rays* are high-energy photons that carry off energy as electromagnetic radiation, requiring a good few inches of lead to prevent them from zipping through you.

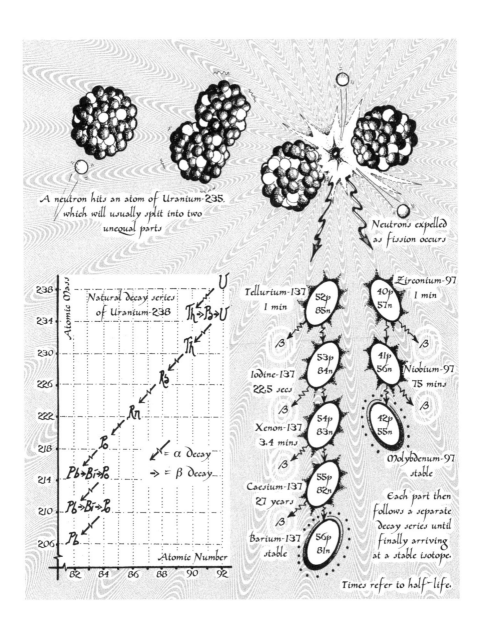

A neutron hits an atom of Uranium-235, which will usually split into two unequal parts

Neutrons expelled as fission occurs

Atomic Mass

238

234

230

226

222

218

214

210

206

Natural decay series of Uranium-238

U

Th→β→U

Th

Ra

Rn

Po

Pb→Bi→Po

Pb→Bi→Po

Pb

↙ = α decay
→ = β decay

Atomic Number

82   84   86   88   90   92

Tellurium-137
1 min
52p
85n

β

Iodine-137
22.5 secs
53p
84n

β

Xenon-137
3.4 mins
54p
83n

β

Caesium-137
27 years
55p
82n

β

Barium-137
stable
56p
81n

Zirconium-97
1 min
40p
57n

β

Niobium-97
75 mins
41p
56n

β

Molybdenum-97
stable
42p
55n

Each part then follows a separate decay series until finally arriving at a stable isotope.

Times refer to half-life.

ORBITAL STRUCTURES
the whirly world of the very small

At the scale of fundamental particles like the electron, energy comes in discrete packets, or *quanta*. Strangely, everything down here behaves like both particles *and/or* waves, depending on your perspective.

If an atom is pumped up with energy, excited electrons whizzing around the nucleus make sudden *quantum* energy jumps into new orbital levels that fit together like a buzzing ethereal flower.

Mathematical *wave functions* can predict the probability of finding an electron in a specific place. Most of time the electron will be within the main part of its density plot (*opposite left*). However there is always a slim chance it could be somewhere else entirely.

Each orbital can be inhabited by two electrons, which need to have opposite spins to each other. The sphere (*opposite top right*) represents the primary *1s* orbital and its twins may be anywhere, including in the nucleus. The second orbital set, the *2p*, fills as shown below. Three double teardrop shapes reflect around a nuclear *nodal plane* where electrons hardly ever go. As further orbitals fill, new electrons are forced into more and more exotic dances (*lower opposite, and page 389*).

Weirdly, the wavy nature of electrons means that we can't know their position *and* speed at the same time. Just looking at something so tiny radically alters its behaviour. A small but measurable *uncertainty* creeps in.

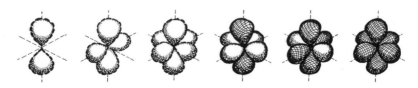

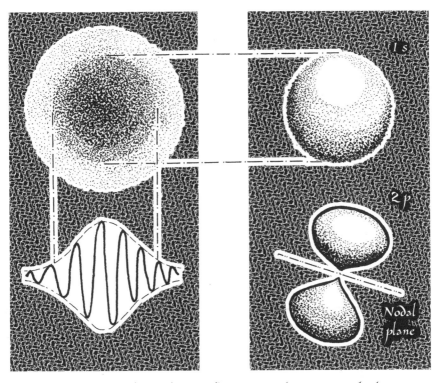

Finding the elusive electron : from wave packet to atomic orbital

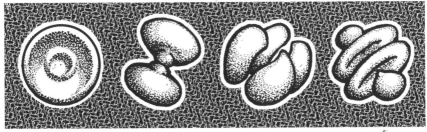

2 s   3 p   4 d   5 f

An assortment of orbital shapes : note that only s orbitals lack a nodal plane

MATERIALS SCIENCE
the future is invisible

Peering into atoms and molecules to understand how they work has enabled the manipulation of matter on finer and finer scales, producing immensely strong alloys for deep ocean and space exploration, 3D printers to assemble things a slice at a time, and fuel-cells and solar panels to keep things going when the oil runs out. Theory can predict new chemistry, new possibilities, and practical offspring (*opposite top*). Novel plastics, polymers, ceramics, and composites help make stuff that is ever more flexible, adhesive, superconducting, and heat resistant.

The children of the quantum revolution can be found in the flashing LEDs of everyday electronics, and the silicon microchips of their brains, all of this due to the erratic way electrons wander through certain crystals. Spin engineering, which uses fundamental properties of atoms and electrons, has allowed us to peek harmlessly into compounds or living tissue by nuclear magnetic resonance. Spintronics will probably form the minds of the next generation of computers (*center opposite*).

Today, we can actually push single atoms about, leading to the science of *nanotechnology* where the odd effects of the quantum realms are harnessed. The size of a nanometer is one billionth of a meter (the length fingernails grow in a second), and applications for nanotech are already present in medicine, electronics, and chemical synthesis (*lower opposite*). One day designer molecules may build themselves, and microscopic intravenous nanobioelectronic robots could rid us of disease, or alas turn all they touch into grey goo. Molecular engineering has huge consequences, and with our environment already stressed, we need to be extremely careful about our powerful new toys.

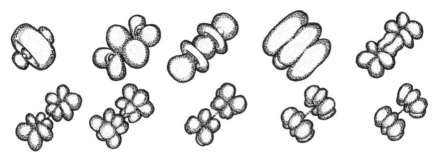

Quintuple Diuranium bonding – using quantum chemistry it has been discovered that two uranium atoms may form molecules held together with five covalent bonds

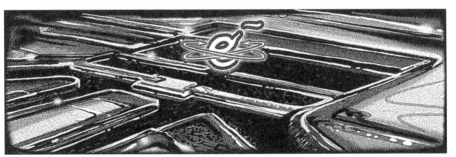

A qubit – at the heart of a quantum computer each qubit is able to hold and measure electrons in a superposition of 0 AND 1 simultaneously. Classical computers can only deal with 0 OR 1

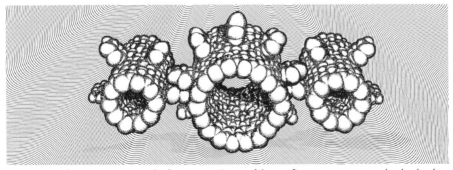

Nanotech gearing – nanotechnology opens the possibilities of engineering at a molecular level

Quirky Quarks
and curious quantum effects

Deep inside the *nucleons* (protons and neutrons) lurks an even smaller realm. Energy and matter down here have such a close relationship that it's sometimes difficult to tell them apart.

Both nucleons are made of three *quarks*, particle-like matter fields that are the building blocks of the everyday universe. The quarks we usually encounter are called *up* and *down*. Peeking inside a proton (*opposite top left*) we find two up quarks and one down quark. Up quarks have an electric charge of $+\frac{2}{3}$, whilst downs carry $-\frac{1}{3}$. Adding the three quark charges gives the proton's total of +1. The neutron is composed of a slightly different crew—two down quarks and one up quark, which cancel out, leaving it electrically neutral (*opposite top right*).

Binding quarks together is the special job of the *nuclear strong force*. Intriguingly, instead of the two charges of electricity, the strong force has three charges to balance which can be likened to the three primary colours of light. Mixing red, green, and blue light together produces a neutral white, and similarly, each nucleon carries three different 'colour' charged quarks which combine into a neutral overall charge. The colour force is carried by peculiar fundamental particles called gluons.

Weird quantum effects rule this deep level of matter. When quarks are pulled apart, pairs of quarks and mirror-image antiquarks seeimgly magically pop into being. Since the strength of the strong force remains constant over its short range, regardless of distance, like unbreakable elastic bands stretched to their limit, it becomes more efficient to borrow energy from the quantum vacuum than to overcome the gluon bonds that hold the quarks together (*lower opposite*).

The Proton
Two Up quarks and one Down
$2/3 + 2/3 - 1/3 = +1$

The Neutron
One Up and two Down
$2/3 - 1/3 - 1/3 = 0$

Quarks exchanging gluons to balance
the red, green, and blue colour charges
of the strong force, overall giving a
neutral charge across the nucleon.

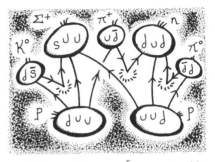

A Feynman diagram of two protons (p)
transforming into a neutron (n), a strange
sigma (Σ^+) particle and a host of mesons from
spontaneous quark/antiquark creation

Confinement is due to the strong force colour charge remaining constant as distance
increases: as quarks are pulled apart, it is energetically easier for quark/antiquark
pairs to be created than to stretch the gluon fields holding them together.
Only very very short lived top quarks have been observed alone

THE FOUR FORCES
holding the universe together

Everything in the universe interacts with everything else through four universal forces, carried by four types of wave-particles called *gauge bosons*.

The *electromagnetic* force is carried by *photons*. Light, x-rays, radio waves, and microwaves are all different frequency wrigglings of electromagnetic fields. This dual-aspect force attracts electrons to protons and causes most annihilations between matter and antimatter (*see below*). It is also veritably the prime mover of all chemical reactions.

Acting over distances the size of a nucleus, the *strong* force binds quarks by exchanging eight types of *gluon*, the carriers of colour charge. The strong force only affects quarks and gluons.

Particle decay is governed by the *weak* force, which acts over extremely short ranges. Changing a down quark into an up one, for example, transforms a neutron into a proton and hence an element into the next in the periodic table. Carried by *W* and *Z vector bosons*, the weak force also allows neutrinos, very light fundamental particles, to perform their rare interactions with everyday matter (*see page 362*).

Gravity, although by far the weakest force, nevertheless operates over almost infinite distances, extending its grip over all matter. Finding a quantum explanation for gravity has had its up and downs, though a carrier boson, the *graviton*, has been tentatively described.

Positron (*Anti-electron*)

Electron

$E = Mc^2$

Photons

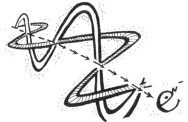

The electromagnetic force can operate over large distances: it shapes electron orbits and hence controls chemical behaviour in atoms

The strong force acts over distances the size of a nucleus, shaping and holding the quarks within together

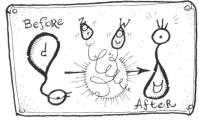

The weak force has a very short range indeed and is responsible for quark transformation and neutrino interactions

Gravity acts across very great distances and connects all types of matter, from galaxies to the atoms of a cat

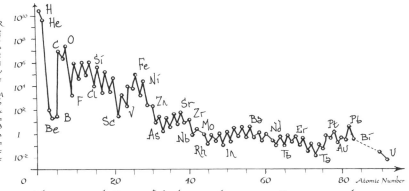

Relative cosmic abundances of the elements: the zig-zag is due to atoms with even numbers of protons and electrons being slightly more stable than ones with odd numbers

Quarks, Leptons, and Mesons
the fundamental families of matter

Everyday matter is made of up and down quarks, and two *lepton* cousins, the electron and neutrino, in a first generation of indivisible fundamental particles (*opposite top*). At higher energies, like cosmic rays, a second, heavier, family of four is found, *charm* and *strange* quarks, with the *muon* and *muon-neutrino*. At extreme energies a third, even more massive, family appears. Each member also has a mirror-image antimatter twin.

All matter has *spin*, a quantum version of angular momentum. Quarks and leptons have spin $\frac{1}{2}$, needing to 'turn' around twice to look the same (*see slugs below*). Particles with half spin= ($\frac{1}{2}, \frac{3}{2}, \frac{5}{2}\dots$) are called *fermions*, bound by the *Pauli exclusion principle*, which forbids two fermions to be in the same quantum state. For example, electron pairs need to have opposite (up or down) spin to share an atomic orbital. *Baryons*, composites of three quarks, like protons and neutrons, are also fermions.

Particles with integer spin (0, 1, 2…) are called *bosons*, the force-carrying gauge bosons and *mesons* (quark/antiquark pairs) being examples (*see appendix page 385*). Bosons can sit in the same quantum state as each other, like photons in a laser beam acting in tandem.

Under certain conditions, even numbers of fermions can act like bosons, giving rise to weird quantum behaviours on a macroscopic scale, like ultra-runny superfluids with no viscosity, or superconductive electrons joining in Cooper pairs to flow without resistance.

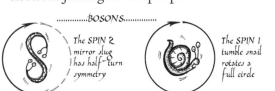

..............BOSONS..............

The SPIN 2 mirror slug has half-turn symmetry

The SPIN 1 tumble snail rotates a full circle

FERMIONS

The SPIN 1/2 moebius slug twists through two circles 720°

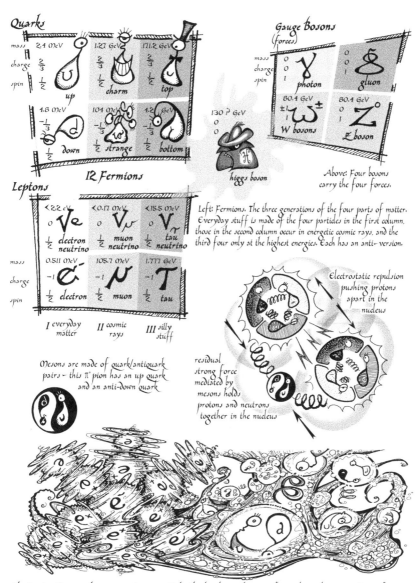

Quarks

mass	2.4 MeV	1.27 GeV	171.2 GeV
charge	$\frac{2}{3}$	$\frac{2}{3}$	$\frac{2}{3}$
spin	$\frac{1}{2}$	$\frac{1}{2}$	$\frac{1}{2}$
	up	charm	top

	4.8 MeV	104 MeV	4.2 GeV
	$-\frac{1}{3}$	$-\frac{1}{3}$	$-\frac{1}{3}$
	$\frac{1}{2}$	$\frac{1}{2}$	$\frac{1}{2}$
	down	strange	bottom

Gauge Bosons
(forces)

mass	0	0
charge	0	0
spin	1	1
	photon	gluon

	80.4 GeV	80.4 GeV
	± 1	0
	1	1
	W bosons $\pm$	Z boson

Above: Four bosons carry the four forces.

130 ? GeV
0
0

higgs boson

12 Fermions

Leptons

	<2.2 eV	<0.17 MeV	<15.5 MeV
	0	0	0
	$\frac{1}{2}$	$\frac{1}{2}$	$\frac{1}{2}$
	electron neutrino	muon neutrino	tau neutrino

mass	0.511 MeV	105.7 MeV	1.777 GeV
charge	-1	-1	-1
spin	$\frac{1}{2}$	$\frac{1}{2}$	$\frac{1}{2}$
	electron	muon	tau

I everyday matter   II cosmic rays   III silly stuff

Left: Fermions. The three generations of the four parts of matter. Everyday stuff is made of the four particles in the first column, those in the second column occur in energetic cosmic rays, and the third four only at the highest energies. Each has an anti-version.

Electrostatic repulsion pushing protons apart in the nucleus

residual strong force mediated by mesons holds protons and neutrons together in the nucleus

Mesons are made of quark/antiquark pairs – this π pion has an up quark and an anti-down quark

The Quantum Vacuum : there is no empty space. At the Planck scale a seething sea of virtual particles pop in and out of existence. An electron surrounded by virtual counterparts (left). Quark/antiquark mesons created by balancing energy debts (right).

EXOTIC PARTICLES
and subatomic siblings

Energetic cosmic rays from outer space, high-speed charged particles like protons and helium nuclei, with a few electrons and tiny amounts of antimatter, constantly hit our planet, ionizing atoms in the upper atmosphere to create cascades of subatomic particles. Visible as the glowing polar auroras, branching streamers of *hadrons* (quark composites of baryons and mesons) shed electrons, positrons, and γ-rays to cause further chain reactions, short-lived exotic particles rapidly decaying into more stable, lower energy ones (*opposite top*).

In experiments re-creating the extreme energies of the early universe, accelerators collide protons and other matter at near light speed to reveal hidden structures. Detectors trace the curling paths of hundreds of colliding and transmuting particles (*below*). Some are higher energy versions, or *resonances*, of others, linked together by symmetrical patterns into families. One, called the *eightfold way* (after a Buddhist doctrine), charts family relationships as octets (*lower opposite*) correlating charges, spins, and other characteristics. The quantum orchestra plays some surprising tunes as nature reveals her subtle harmonics.

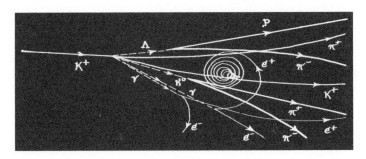

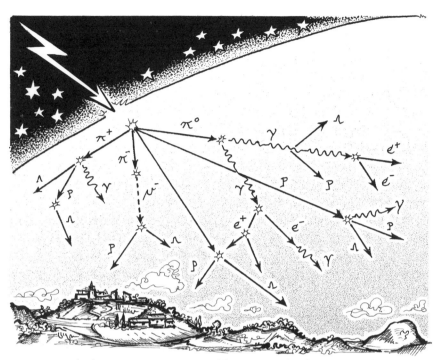

A high energy cosmic ray sets off a cascade of subatomic particles

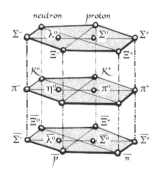

Eightfold way octets

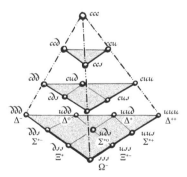

Spin 3/2 baryon family tree

173

Quantum Theories

interpretations and entanglements

To this day no one understands *why* the application of matrix mechanics in the complex plane predicts electron orbitals (*see pages 112–117*), nor *how* an entangled pair of photons seem to be able to communicate instantly regardless of how far apart they are separated (*below*), nor *why*, in the twin slit experiment, a single fired photon, atom, or molecule interferes as a *wave* (so passes through both slits) when unobserved, yet behaves as a *particle* (going through just one slit) when observed (*opposite top*).

There are different *formulations* of quantum mechanics. All work beautifully, making precise predictions for building our mobile phones and computers, but each leads to a different *interpretation* of what is happening, and so the true nature of reality. The most widely taught is the *Copenhagen interpretation*, which holds that microscopic reality is *actually* created by observation—*looking* causes the collapse of the probabilistic wave function. Next up is the *Many Worlds interpretation*, which says that the world constantly divides, so the moment the cat looks in the box he really does create one world in which the scientist is alive and another one in which he is dead (*lowest opposite*). Another, the *Transactional interpretation* allows the future to affect the past, while the *Bohm interpretation* sees the entire universe as a single entangled whole. There are many more. The truth is that we do not know, but we do know that the future is quantum, and that it's going to be different.

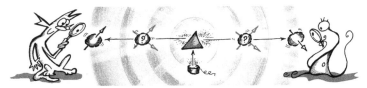

The Twin Slit experiment:
Wave or particle? Unobserved,
single photons passing through
two slits produce an interference
pattern, like waves would. however
if we try to see which slit the photon
went through, the pattern disappears
as if the photons were discrete particles

The Uncertainty Principle in action: An electron is trapped in
an energy well. Its wave function though extends beyond the
barrier giving a probability of finding it on the other side.
This quantum tunnelling is the cornerstone of electronic
components like diodes, transistors, and silicon chips

The Casimir Effect: Very close parallel
conducting plates are pushed together
by fluctuations in the quantum vacuum.
Since only electromagnetic waves with
resonant wavelengths can fit in the
cavity, the field pressure between the
plates is lower inside than that outside

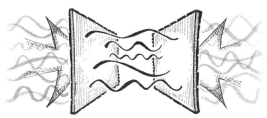

Kat and Schrödinger: A scientist is in a closed
box with a quantum system that gives a 50/50
chance of survival. The probability wave function
says that the system exists in a superposition of
states until observed where the hapless subject is
both alive and dead. Only when measured will
the wave function seem to decohere to one state.

Strings and Things
bubbles and branes

The more we probe into the wispy knottings of wave-matter, the more the colossal energies bound up in the subatomic realm present a field day for adventurous mathematicians. Many attempts to find a *Theory of Everything* invoke additional dimensions to the usual four of spacetime. Quantum models that include gravity need eleven or so dimensions to work, and use *supersymmetries* to relate fundamental families and forces to phenomena. Most envisage an initial singularity connected through a single unified force, folded and condensed into the familiar four.

The knitting of *superstrings*, hypothetical one-dimensional standing wave threads which loop and resonate, along with the lure of *supergravity* are different views over a much larger vista, the mysterious M-theory. This pictures the universe as ripples in an infinitely vast and thin *p-brane*, a membrane spanning many dimensions stretching through hyperspace. Was the Big Bang branes colliding, our little universe being merely the interference patterns where they crossed? Others theories suggest that we are just one universe amongst a multitude, or possibly a 3D projection cast from a 2D information horizon.

Elegant geometries perhaps underlie the inner workings of reality, with particles, spins, and forces all surfing around a deeper symmetry. It could be that any description of nature will of necessity remain incomplete, the subtleties of the quantum realm obscuring attempts to peer too far into nature's infinitely layered onion. Formed from hidden harmonies, could we be only as real as the holographic sparkles in a sunny pool, or the fractal rainbow swirls on the surface of a soap bubble?

177

BOOK IV

EVOLUTION

Gerard Cheshire

with additional content by Lindi Houseman

Introduction

There are few peoples on Earth who do not have a creation myth. The Native American Iroquois believe the world and everything on it was created by Sky People, the ancient Japanese believed the world was created by gods who grew from a single green shoot, and many people today still fervently believe that the universe and all life in it was created in one form or another by a god.

This little book tells the remarkable tale of a modern creation story that has been meticulously pieced together over the last 150 years by hundreds of thousands of botanists, zoologists, chemists, and biologists working in jungles, fields, zoos, oceans, and laboratories all across the world. Instead of the rich symbolism of myth or the rote certitudes of religious doctrine, it is generally couched in the difficult language of experimental science. Still as terrifying to as many people today as it was when first publicized by Charles Darwin in 1859, it tells the unlikely tale of a bacterium that became a kind of worm, that became a fish, that became a reptile, which became a sort of rodent, which became an ape, who became a human, who left Africa, and became you.

Like many creation myths it sounds fantastic. Like all good stories it is full of sex, death, family struggles, kindness, and friendship. It is a story some people have only just heard for the first time, others never before, because we are only now filling in the details. And yet the story is not finished at all, it is still unfolding, still evolving, still being told.

If our species survives the era of mass extinctions we are currently manufacturing for our fellow travellers on this little life-crusted ball of fire we all share, we will no doubt become other things yet.

LIFE'S GREAT FAMILY
the fog rises

Amidst the foggy ideas of past centuries there arose occasional glimpses of a strange new notion: that humanity, along with all other organisms, instead of being created outright, had instead arisen through a process of biological adaptation—*evolution*.

Publishing *Systema Naturæ* in 1735, Carl Linnaeus (1707–78) replaced the classical categorization of animals by their mode of movement, with the system of *kingdoms, phyla, classes, orders, families, genera,* and *species* still used today. It seemed evident that these families of animals and plants had evolved in some way from common ancestors, or from one another, and by the 1800s scientists were trying to work out exactly how. In 1809 Jean-Baptiste Lamarck (1744–1829) proposed that species evolved via acquired characteristics, so that subtle (and often useful) changes made to their design during their own lifetimes (e.g. a tennis player's better developed arm muscles) were passed on to their offspring. This theory, however, although popular, had serious flaws. It turned out that offspring often varied wildly from their parents, and, importantly, that characteristics acquired over a lifetime, such as injuries or larger muscles, could not be passed down the generations either.

The theory was not working. Something was missing.

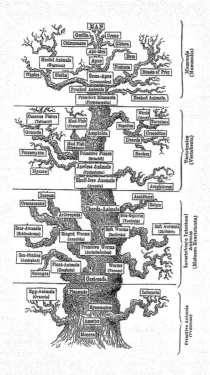

Left: An early Linnaean tree of life, the kingdoms shown as a hierarchy with mammals at the top, and man at the summit. Although a breakthrough, these early versions were not so very far from the medieval Chain of Being, the hierarchy of souls, with God at the top, then angels, humans, animals, vegetables, and finally minerals, each kingdom having natural authority over and control of those beneath it.

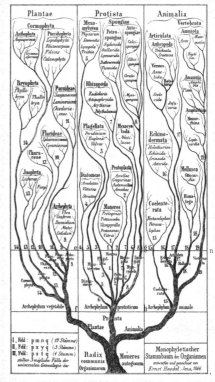

Right: Ernst Haeckel's original tree of life, drawn in 1866, dividing living things into three basic groups, plants, animals, and protists (a diverse group of eukaryotes, or multicellular organisms, which do not fit into the other eukaryotic kingdoms). Haeckel coined the term 'protist' for this diagram. Modern classifications vary from this diagram in certain important ways (e.g. fungi are today considered a kingdom of their own). To see a more modern version of this tree of life diagram turn to page 227.

THE BIG IDEA
eat, breed, adapt, and pass it on

In 1859, after over 25 years collecting specimens and studying variations between species, particularly barnacles, Charles Darwin (1809–82) revealed his theory of evolution by natural selection. It directly contrasted with Lamarck's theory. The fact that offspring vary in their characteristics, suggested Darwin, was enough to allow nature to select in favour of those individuals slightly better suited to perpetually changing environments. Tiny changes, each conferring small advantages, could mount up, over many generations, to produce large differences, even new species. In 1864 Herbert Spencer (1820–1903) coined the phrase 'survival of the fittest,' attempting to encapsulate this idea.

Darwinism usurped Lamarckism, though no one, including Darwin, was yet able to furnish an empirically evidenced explanation for the mechanism that promoted the variation that allowed natural selection to work. Nevertheless, *gemmules*, his descriptive term for the particles responsible for biological transfer, were, unknown to him at the time, in many respects similar to Mendel's pairs (*see page 190*).

Darwin's theory also suggested that man had descended from ancestral apes. A revolutionary concept at the time, his theory of evolution questioned humanity's place in the universe and challenged long-held beliefs about the nature of creation.

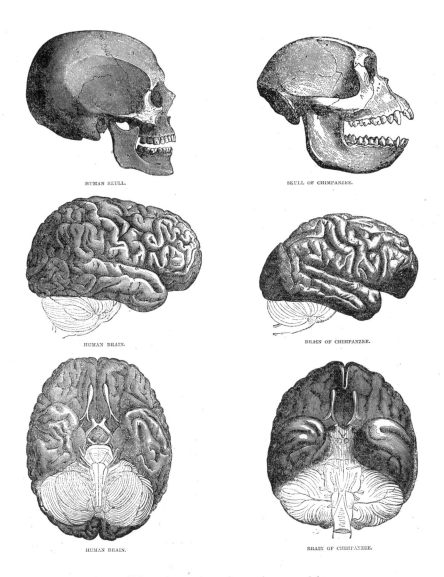

HUMAN SKULL.

SKULL OF CHIMPANZEE.

HUMAN BRAIN.

BRAIN OF CHIMPANZEE.

HUMAN BRAIN.

BRAIN OF CHIMPANZEE.

Above and left: Striking similarities between humans and chimpanzees
suggested that both were closely related, and that man was a species of ape.
No piece of evidence to the contrary has ever been found since, only further support.

LIVING PROOF
and dead ends

In support of his evolutionary theory Darwin found important examples of evolution in action. One was *artificial selection*, or *eugenics*. Darwin argued that humans had created domesticated plants and animals by imposing selective processes on captive populations. Careful breeding, he suggested, created desirable characteristics in dogs, cats, horses, pigeons, and chickens (*opposite top*) in much the same way that nature did in the wild.

Travelling on the *Beagle* (1831–36), Darwin had noticed groups of closely related species that appeared to have adapted to slightly different environmental demands. Examining the exotic reptiles and birds of the Galapagos archipelago in 1835, he found that each island had its own idiosyncratic species of tortoise (*opposite*) and finch (*below*), demonstrating that isolation events had allowed natural selection to take populations in different evolutionary directions on different islands from a common ancestor.

Two problems remained: Firstly, Darwin had only demonstrated *lateral* evolution, not *longitudinal* evolution; species could adapt and vary on a theme, but a tortoise appeared to remain a tortoise and a bird a bird, so his theory didn't yet explain how wholly new types of animal and plant came about. Secondly, he lacked a provable mechanism behind this variation.

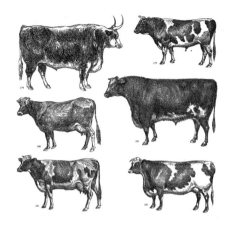

Above: Selective human breeding of cattle has created many hundreds of breeds, each with specific qualities. Some are bred for milk, others for meat, some for hot climates, others for freezing hillsides. Chickens too are selectively bred, some for eggs, others for meat. Darwin bred pigeons at his home to better understand the process of artificial selection and get a first-hand grasp of how quickly small variations in individuals could be passed into entire populations.

Left: Giant tortoises on the Galapagos Islands. There are 11 existing species of tortoise spread over the islands, all probably descended from a single ancestor. On dry islands, where little grows except cacti, taller tortoises with longer necks had the advantage over their shorter counterparts. Taller cacti also tended to survive being eaten and are found on these islands, an example of an evolutionary arms race. The tortoises allow the finches on the islands (far left) to peck ticks from their skin, which suits both parties well, and is an example of a mutualistic symbiotic relationship.

189

THE UNSUNG MONK
peas and their peculiar traits

While Darwin was pondering over his mechanism, a Moravian monk, Gregor Mendel (1822–84), had already been experimenting with heredity for years. Starting in 1856, Mendel had begun breeding pea plants on a hunch that inheritance was mathematically predictable. By 1865 he had tested over 29,000 plants and amassed enough evidence to show that it was possible to accurately forecast the ratios between paired traits (such as smooth vs wrinkled peas, or tall vs dwarf plants) after controlled crossbreeding. For example, crossing purebred tall and dwarf pea plants produced only tall specimens. However, crossing these with each other again, dwarfness reappeared in the next generation, with tall:dwarf plants in a 3 : 1 ratio. Mendel concluded that pairs of particles (now known as *alleles*), one *dominant* and one *recessive*, were at work (*see example, opposite top*).

Mendel was right. We also now know that other species, such as snapdragon flowers, can also display *incomplete dominance when red and white varieties are crossed* (*lower opposite*). And there is *codominance*, where neither allele is recessive. An example is the ABO blood-group system, which is controlled by three alleles, A, B and o. o is recessive to both A and B, and causes O-type blood, while A and B are codominant. You inherit two alleles, one from each parent, and can therefore either end up with A(*AA, Ao*), B(*BB, Bo*), AB(*AB*), or O(*oo*) blood groups. We also now know that just one letter change on chromosome 9 makes the difference between O and A, but more of that to come.

Darwin didn't hear of Mendel's work and it was not fully recognized until 1900 by William Bateson (1861–1926).

DOMINANT and RECESSIVE TRAITS

TT   dd
purebred
Tall and dwarf peas

crossbred
produce
only tall

Td   Td
first generation (children)

crossbred produce 3:1 tall:dwarf

Above: Mendel's original experiment on peas. If the original purebred plants' pairs of particles were T-T (tall, dominant) and d-d (dwarf, recessive), then the second generation would all be T-d, all tall, while in the third generation equal numbers of T-T, T-d, d-T, and d-d would, because of the dominance of the T, produce the observed 3:1 ratio of tall to dwarf peas.

TT   Td   Td   dd
second generation (grandchildren)

INCOMPLETE DOMINANCE

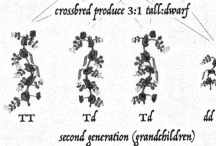

RR   ww
purebred
Red and white snapdragons

crossbred
produce
only pink

Rw   Rw
first generation (children)

crossbred produce 1:2:1 red:pink:white

Above: An example of incomplete dominance, here in snapdragons. The original plants are R-R (red, partially dominant) and w-w (white, recessive). The second generation are all R-w, pink, while the third generation show red, pink and white children in the ratio 1:2:1.

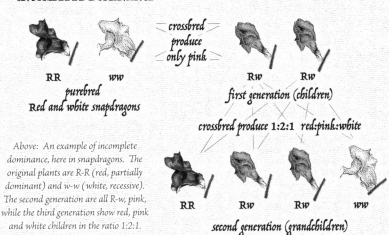

RR   Rw   Rw   ww
second generation (grandchildren)

CHROMOSOMES
genes and DNA

Toward the end of the nineteenth century, scientists began turning their microscopes on cell nuclei, looking for the components responsible for the evolutionary mechanism, and coining the term *chromosome* for the stripey pillules seen in the nucleus. Observations of cell division (*mitosis*), gamete production (*meiosis*), and fertilization showed that chromosomes behaved in an organized way and it was soon suggested that they might carry inherited information as strings of heredity particles. By the 1920s the black filaments inside chromosomes were revealed as chains of base/sugar/phosphate nucleotides, *deoxyribonucleic acid*, or 'DNA'. Its double-helix structure was discovered in 1953.

DNA is a 4-letter code for life, universal to all life on Earth, i.e. all organisms use it in exactly the same way. The number of chromosomes varies from species to species (*opposite top*), but all animals carry two versions of each chromosome in every cell nucleus, one from Mother, one from Father, and spaced out along every chromosome are special sections of DNA called *genes*. Other exclusively maternal DNA is found outside the nucleus, in mitochondria, the cellular battery packs.

The 23 chromosomes of the human genome. Two copies of each are found in every nucleus. One comes from your father, one from your mother. A Y from father makes you male.

192

ANIMALS

3 MOSQUITO 6
4 DROSOPHILA 8
6 HOUSE FLY 12
12 SALAMANDER 24
13 LEOPARD FROG 26
16 ALLIGATOR 32
20 SHREW 40
20 SQUIRREL 40
22 BAT 44
22 PORPOISE 44
23 HUMAN 46
27 GARDEN SNAIL 54
28 ELEPHANT 56
30 GOAT 60
32 ARMADILLO 64

32 GUINEAU PIG 64
32 OPOSSUM 64
32 PORCUPINE 64
35 CAMEL 70
37 CHICKEN 74
39 DOG 78
41 TURKEY 82
66 KINGFISHER 132
104 KING CRAB 208

PLANTS

7 PETUNIA, X2, 14
7 PEA, X2, 14
7 LENTIL, X2, 14
7 RYE, X2, 14
7 EINKORN WHEAT, X2, 14

7 DURUM WHEAT, X4, 28
7 BREAD WHEAT, X6, 42
8 ALFALFA, X4, 32
9 LETTUCE, X2, 18
10 CORN, X2, 20
11 BEAN, X2, 22
11 MUNGBEAN, X2, 22
12 POTATO, X4, 48
12 TOMATO, X2, 24
12 RICE, X2, 24
12 PEPPER, X2, 24
14 APPLE, X2, 34
14, BRAMLEY APPLE, X3, 52
20 SOYBEAN, X2, 40
24 TOBACCO, X2, 48
41 LILY, X2, 82
630 FERN, X2, 1260

Above: Chromosome numbers in certain animals and plants. All animals are diploid, which is to say they carry two copies of each chromosome in every cell nucleus. For example, bats have paternal and maternal copies of 22 chromosomes, so 44 in each cell. Plants can be polyploid, having more than two copies, so triploid (three copies, normally infertile crossbreeds), tetraploid (four copies), or, even, hexaploid.

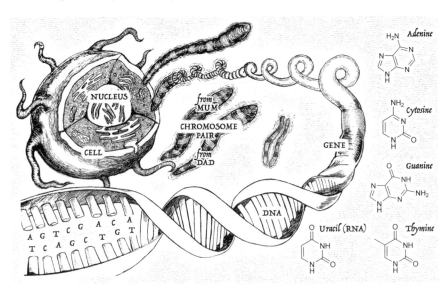

Above: Chromosomes live in the nucleus and are made of DNA. The double helix structure, like a twisted ladder, is comprised of pairs of just four bases, adenine (A) always bonding with thymine (T), and guanine (G) with cytosine (C). DNA thus stores information in a double-binary or quadrinary fashion, with each of the two strands being an identical copy of the other. A few thousand genes are spaced out along each chromosome, with long sections of repetitive non-coding DNA creating space between them.

THE BOOK OF LIFE
four letters, twenty words

The genome of a species is the entire DNA sequence found in its chromosomes. The human genome is like a cookbook as long as 1,000 bibles, with 23 chapters (chromosomes), each chapter containing several thousand recipes (*genes*). Each recipe is for one protein, and is written using just 20 different words (*codons*), made of only four letters (*bases*). The recipes have advertisements (*introns*) in them, which have to be snipped out for the finished copy (*exons*).

When the human genome was finally mapped, scientists were surprised to find it contained pages and pages of gobbledygook *between* each gene. Some sections of these non-coding (or 'junk') DNA sequences originate from distantly broken genes, while others are repetitive transcription errors (DNA can lose count when copying repetitive sequences like CATCATCAT). Other sections are dead retroviruses (viruses which use reverse transcriptase to copy their RNA into their host's DNA, so that they become part of their host's genome).

Another group of genetic parasites, descended from retroviruses, are known as 'jumping genes'. These little sequences are found as introns in almost every gene and shout "copy me everywhere" to passing chemical equipment. Highly virulent little data bugs, today they make up about a quarter of our DNA. 'Real' genes account for only 3%.

Non-coding DNA does, however, have its uses. It creates spaces between genes, aiding clear transcription, and preventing their breaking during *crossover* (*see page 196*). Also, though not expressed as proteins, non-coding DNA can also finely regulate genetic expression, subtlely enhancing, modulating, or suppressing the transcription and therefore the expression of nearby genes, a form of *epigenetics*.

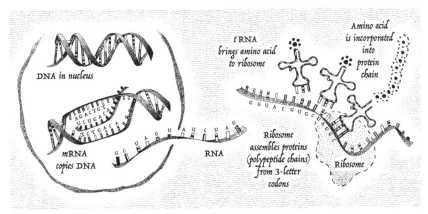

Above: DNA in the nucleus comprises of adenine (A) to thymine (T) and guanine (G) to cytosine (C) bonds. It uncoils and is transcribed into a strand of messenger RNA, identical to DNA except for the replacement of thymine (T) by uracil (U). The transcription unit is read in three-letter words, each of which code for an amino acid. These are then converted into a string of amino acids known as a polypeptide chain, or protein.

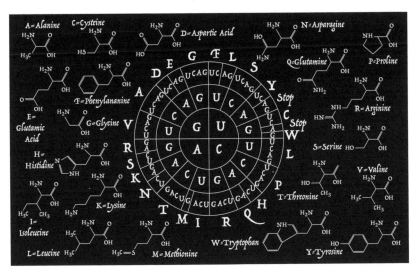

Above: The code of life is written with just four letters and words (codons) of three letters. Use the chart above, starting at the center, to find out which amino acid is produced by any word. For example UAC leads to Y, Tyrosine. Most amino acids are coded for by more than one word, so that there are only 20 amino acids used by life. All life on Earth, whether tree, beetle, rabbit, human, or fungus, uses exactly the same code.

A WORLD OF VARIATION
how it all gets mixed up

Darwin's theory depended on a mechanism that could create small variations in offspring. The answer was hidden deep within the way *gametes* (sperm, eggs, or plant spores) are produced.

As Mendel had guessed, a complete copy of all the DNA needed to build an entire organism is held in every cell nucleus, organized into discrete chromosomes, with one *homologous* (similarly mapped but different) copy from each parent. During *meiosis* (*opposite*), genes are snipped out and swapped between the parental pairs, the reshuffling and recombining creating new chromosomes for the gametes, each with different trait potentials (*crossover shown below*).

This was one source of the variation Darwin had been looking for. Indeed, shuffling is all very well, but wild cards also play their part. During meiosis all sorts of things can go awry: copying errors, occasional deletions, duplications, sometimes inversions of sections of DNA, sometimes in genes, but more often in non-coding DNA. Often subtle (the tiniest change to a single letter or protein), they occasionally turn out to be advantageous, but can also be fatal.

For instance, there is a gene on chromosome 4 which simply contains the word 'CAG', repeated again and again. Most people have it repeated from anywhere between 6 and 30 times, but if you have it repeated over 35 times you will slowly die of Wolf-Hirschhorn syndrome. Or another example: A single-letter change in a 253-word gene on chromosome 20 will give you mad cow disease.

MITOSIS = CELL DIVISION

PROPHASE
DNA in nucleus.
Aster and spindles
develop from centrioles.

PROMETAPHASE
Nuclear envelope
breaks down.

chromosome/chromatid
pairs (exact copies of
chromosomes) attach to
spindles.

METAPHASE
Chromosome/chromatid
pairs line up at the
equator of the cell.

ANAPHASE
The spindles pull
the chromosomes and
chromatids apart.

In humans, 46
chromosomes are pulled
into each half.

TELOPHASE
chromosomes reach
mitotic poles and
cell starts to pinch.

CELL
DIVISION
followed by

INTERPHASE: chromatids created for next division.

MEIOSIS = PRODUCTION OF SPERM OR EGG

PROPHASE I
Maternal and paternal
copies of chromosomes
pair up.

Each chromosome
replicates, creating a
chromatid.

CROSSOVER occurs
between maternal and
paternal versions

METAPHASE I
Shuffled chromosome/
chromatid pairs
are separated

TELOPHASE I
Division
into two

PROPHASE II
Division
METAPHASE II

ANAPHASE II
Further
separation of pairs

TELOPHASE II
Creation of four
haploid cells, each
with one copy of
each chromosome

Nurturing Nature
the strange behavioural sieve

In 1896 James Mark Baldwin (1861–1934) advanced a theory that learned advantageous behaviours could eventually become instincts. He proposed that behavioural, cultural, and even chemical factors could greatly shape a genome. His ideas were ahead of his time, as he had effectively predicted the modern study of memetics (*see page 256*) and epigenetics (*see page 202*).

Organisms possess genes that incline them to behave in certain ways (and also in certain ways at certain times). When circumstances change, which they invariably do, individuals more inclined to acquire appropriate new behaviour patterns will survive better, breed more, and amplify those inclinations (*see opposite*).

The debate between those who believe that habits are primarily dictated by DNA and those who believe that they are mostly learned has become ever more tangled. Today, many instincts are understood as needing to be triggered *via* appropriate nurture. For instance, a monkey reared in the wild learns the fear of snakes from its mother (perhaps she screamed on seeing them). A young monkey reared with no knowledge of snakes quickly learns to be scared of them if adults communicate fear in the presence of one. But what is interesting is when an adult monkey which has been trained to be scared of flowers (don't ask how) is introduced to a young monkey reared in captivity, he can scream at flowers all day but the young monkey will just look at him as if he is crazy. It turns out there is no innate triggerable fear of flowers, whereas there is one for snakes. Many genetic instincts work like this, and lie dormant until switched on or accessed. Only certain patterns at the right time will cause them to come to life.

Above: A group of frogs, recently hatched from a small pool, attempt to cross a river. Different frogs have different inclinations, causing them to learn to either cross a log, jump lily pads, or head for the branches overhead.

Above: Natural selection takes its course. The slight differences in behavioural inclinations have favoured the frogs that learned to climb trees. All the others fall victim to predators in the river. The tree frogs get to breed.

Above: The useful behavioural inclinations spread through the gene pool and become instincts. New behavioural inclinations and other examples of variation begin to appear and the process begins again.

THE EVOLUTION OF SEX
and now we're stuck with it

One of the hottest topics of debate amongst evolutionary scientists is how and why sex started—it troubled Darwin's grandfather, and Aristotle mused on it. It is now established that sex initiated in protists (*see page 228*), probably in the Mesoproterozoic era, and involved fusion of at least two organisms, whose union then managed to replicate as a viable life-form. All species that reproduce sexually, which is anything complicated, including most plants, fungi, and animals, still revert to a protistal form as gametes (*page 272*). A sperm is essentially a swimming haploid protist with its unmistakable 9:2 microtubule flagellar structure.

Apart from that small clue it's anyone's guess. Sex may have started as a viral-style parasitism for self-replication, or as a mutualistic DNA repair mechanism, or in semi-cannibalism, echoes of which it retains in creatures like the praying mantis (whose female eats the male after copulation) and cuttlefish (in which males lose their whole penis during sex and appear to suffer from post-coital depression).

Sex looks costly in bioenergetic terms—only half the species can gestate, and mating behaviour is an expensive business, even after you've found one. But the advantages are manifold. Genetically it compensates for DNA damage, phenotypically it favours positive features over generations, and by the creation of variation and novelty, it accelerates evolution, for example in antigenic wars with parasites (*page 204*). Rapid evolution is crucial when things are changing, but once a successful design has been achieved, such as by crocodiles (which have changed little for tens of millions of years), reverting to efficient cloning is not an option, and species are stuck with sex forever.

Rotifers can reproduce sexually or
by cloning. Males are smaller, have
no gut, and live only to copulate.
Bdelloids have lost males altogether.

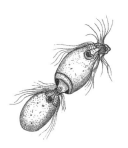

Two protists trying to decide
whether to eat each other or have
sex. Sex may have begun as a sort
of abortive cannibalism in protists.

Hemp, like holly, willows, and
poplars, but unlike most plants,
is dioecious, meaning individual
plants are either male or female.

Nearly all insects mate in classic
animal fashion. Some species like
aphids can clone themselves from
females for a population boom.

Snails are hermaphrodites, each with
a penis and vagina near their head.
They penetrate each other and fire
hormonal love darts during mating.

A male frog chorus is followed by an
amplexus grip. External fertilization
of eggs occurs with non-penetrative
proximity to females.

Clownfish are serial hermaphrodites,
with just one dominant female in a
shoal and a hierarchy among the males.
The alpha-male becomes female.

Seahorse courtship lasts several days
before the female deposits her eggs
into the male. He gestates them for
a few weeks and then gives birth.

Rabbit courtship and copulation
lasts about 30 seconds, and a single
female can generate several hundred
great-grandchildren in one year.

EPIGENETICS
expressing the same thing in different ways

In 1942 Conrad Hal Waddington (1905–75) described a new science, *epigenetics*—a branch of biology that studies those causal interactions between genes and their products which bring the phenotype into being. The term today refers to heritable traits which can be passed down from parents as *expressions* of certain genes, using a process known as methylation, where small markers stick to the parental DNA. Despite there being a full copy of parental DNA in every nucleus, only a small proportion of genes are active, or switched 'on' in any cell at any one time. The bloodstream is full of millions of different chemical messengers that target specific genes in specific cells. By noticing and remembering which genes were 'on' or 'off', sperm and eggs can pass on certain acquired parental traits. A mechanism for Lamarckism.

Environmental conditions, diet and pollution, have all been shown to influence genomic changes in offspring. What you do today really does influence the genome of your great-great grandchild. The epigenetic process has been likened to pieces of chewing gum being stuck on the DNA switches. The gum can be either 'on' or 'off' at any time, inhibiting the switching and expression of the gene. If both gene and gum are 'on', the gene has to stay 'on' until the gum is removed (by an external influence); and vice versa, and in combination.

Sticky gum epigenetics shows how emotions, fears, addictions, and other powerful triggers of hormonal surges (which all course as familiar chemical cocktails through your arteries) can be passed straight on to your children. Additionally, it suggests you really can affect the *expression* of your DNA simply by thinking about things!

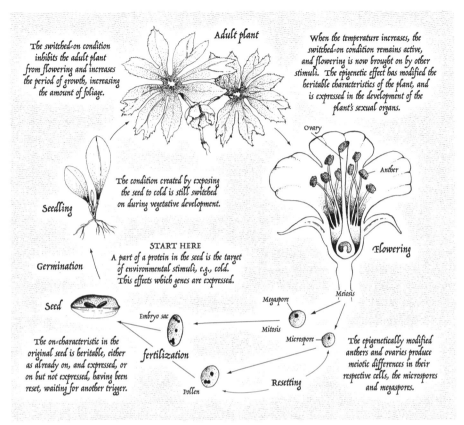

Adult plant

The switched-on condition inhibits the adult plant from flowering and increases the period of growth, increasing the amount of foliage.

When the temperature increases, the switched-on condition remains active, and flowering is now brought on by other stimuli. The epigenetic effect has modified the heritable characteristics of the plant, and is expressed in the development of the plant's sexual organs.

Ovary

Anther

Seedling

The condition created by exposing the seed to cold is still switched on during vegetative development.

Flowering

Germination

START HERE
A part of a protein in the seed is the target of environmental stimuli, e.g., cold. This effects which genes are expressed.

Megaspore

Meiosis

Seed

Embryo sac

Mitosis

Microspore

fertilization

The on-characteristic in the original seed is heritable, either as already on, and expressed, or on but not expressed, having been reset, waiting for another trigger.

Pollen

Resetting

The epigenetically modified anthers and ovaries produce meiotic differences in their respective cells, the microspores and megaspores.

Above: Epigenetics is the study of gene expression. Genes may be unalterable during a lifetime, but their particular expression is affected by many factors and can be passed on to descendants. Genes can be switched on, or off, depending on the cocktail of proteins and hormones flooding the system, and their sensitivity to the cocktail. In the example above, the plant's chemical response to exposing the seed to cold can be passed on to its own offspring.

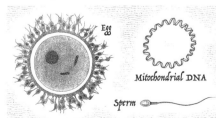

Egg

Mitochondrial DNA

Sperm

Left: Although a father delivers his DNA to his children, only the nuclear DNA makes it through. The DNA for your mitochondria (the bacteriological batteries in your cells), and for the chloroplasts in plants, as well as the chemical soup of intracellular fluid, the vacuoles, and much of the garden that surrounds and supports the nucleus are all inherited only from your mother. Thus mothers can pass on more epigenetic triggers.

THE RED QUEEN
evolutionary arms race

All species are in constant competition with others for resources, and one result of this is that they all have to keep evolving just to *maintain* the status quo. In species that relate as predator and prey, sharper teeth or greater speed in a predator may result in thicker armor or faster legs in its prey. The concept was first described by Leigh Van Valen in 1976, and termed the *Red Queen* effect, after Lewis Carroll's *Through the Looking Glass*, when the Red Queen remarks to Alice, "… it takes all the running you can do, to keep in the same place." It turns out that perpetual motion is a prerequisite of evolution—because environmental conditions are always in flux, so too are the organisms that populate them (*see examples opposite*).

Take the role of sex in fighting disease. Diseases break into cells, and either eat them (fungi and bacteria) or take over their genetic machinery (viruses). They get in by using protein keys, and successful break-ins often lead to the key spreading fast. Sex, as opposed to cloning, creates children who are different to one another and have a variety of different locks to keep the parasites guessing (for example the flax plant has 27 versions of 5 different genes that help resist rust fungus, with different individuals having different combinations). So resistance genes that work well become widespread, but then so do the parasites that unlock them, the corresponding new resistance genes, then the new keys, and so on.

The pace of evolutionary change varies. *Saltation* emphasizes the role of mutations in the sudden morphological changes that result in branching of the evolutionary tree, while *gradualism* emphasizes natural selection and the subtle adaptation of species over time.

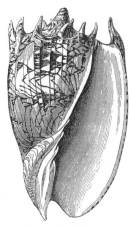

Above: Many predator/prey systems engage in arms races. For example, over millions of years many molluscs have evolved thick shells and spines to avoid being eaten by animals such as crabs and fish. These predators have, in turn, evolved powerful claws and jaws that compensate for the snails' thick shells and spines.

Left: In the arms race between plants and insects, plants that evolve a chemical that is repellent or harmful to insects will be favoured by natural selection. But the spread of this gene puts pressure on the insect population, favouring insects that evolve the ability to overcome this defense. This, in turn, puts pressure on the plant population, and any plant that evolves a stronger chemical defense will be favoured. This, in turn, puts more pressure on the insect population ... and so on. The levels of defense and counterdefense perpetually escalate, without either side ever 'winning'.

SPECIATION

don't play with those children, Rose

When groups of a species are separated and evolve in different directions so far that they can no longer breed with one another, scientists talk of *speciation*, the creation of new species. Speciation is often precipitated by the isolation of and readaptation of a part of the population. In the case of humans specific events probably created a selection pressure. Around 5 million years ago two short ancestral ape chromosomes fused together into a new chromosome 2 in a single individual in an isolated group (*below*). Chimpanzees today still have the original 24 chromosomes.

Darwin struggled to explain how new species evolved from a single ancestral species when no physical isolation or barrier was involved (*allopatric* speciation). In fact, differing behaviour and inclinations in subpopulations can be all that is required for them to genetically isolate themselves. A useful model is that of cichlid fish. In a population frequenting two favourable habitats separated by a barren area, fish inclined to remain within the favourable habitats stand a better chance of survival. Over time natural selection takes the two populations in different directions until they become subspecies and then entirely separate species, unable to interbreed.

Speciation caused by isolation. Opposite: The ape ancestor of humans becomes separated by a huge rift. A single mutation in the isolated group fuses two chromosomes together (6 million years ago) resulting in the creation of a new species. Above: Chimpanzee ancestors play together across a young Congo River. The Congo widens, separating two groups which evolve into modern tool-wielding common chimpanzees and the sex-obsessed Bonobo chimpanzees.

Above: Speciation in a pond. Fish evolve that prefer living under lily leaves. A new area of lilies appears across the pond. The fish show no interest in crossing the barren area between the two areas, so each evolves separately.

Above: Speciation in a mixed population. A varied species. Large, small, light, and dark fish begin to develop sexual preferences for their own size and colour. Eventually two distinct species evolve.

THE MIGRATION OF GENES
out of Africa

Changes in widely spread out gene pools often begin in areas isolated through physical or behavioural causes, localized subpopulations becoming hot-spots of new genetic information. Genes also drift about within species, as individuals explore, travel, and fall in love.

The evolutionary history of humanity shows combinations of both hot-spotting and ubiquitous drift in the fossil record, smoothly progressing in some periods, more staggered in others. Hot-spotting has resulted in the many races of humans, while ubiquitous drift within those races has resulted in relatively uniform genetic characteristics.

There used to be various kinds of tool-fashioning Homo sapiens. Neanderthal man, for instance, was scraping hides in Europe for 500,000 years before mysteriously disappearing in Asia 50,000 years ago, and in Europe about 30,000 years ago, not long after modern trading humans suddenly appeared on the scene, arriving out of Africa.

The study of genes has enabled various maps to be plotted. For example, circular mitochondrial DNA is only inherited from mothers, so avoids the shuffling of meiosis, and remains virtually unchanged down the generations. Studies of mitochondrial DNA have shown that 99% of Europeans are descended from just seven women (called *clan mothers*) living at different sites in Europe at various times during the last Ice Age. Globally, all humans seem to be descended from a single ancestress living in Africa about 200,000 years ago. Similar studies of the Y-chromosome, which is only passed from father to son (also virtually unchanged), have revealed that 99% of Europeans are descended from five men (*clan fathers*) living during the last Ice Age. All humans alive today descend from one man, living in Africa about 70,000 years ago.

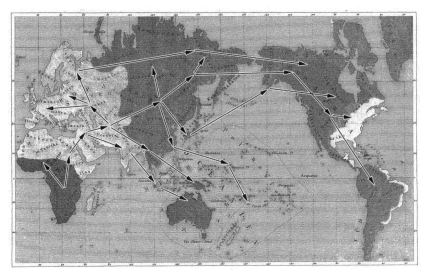

Above: Mitochondrial DNA and the Y-chromosome have revealed the path by which modern Cro-Magnon humans left Africa, by a northeasterly route, roughly 75,000 years ago. Our ancestors split up, heading south, east, then northwest. Europe was colonized around 40,000 years ago, the Americas only 25,000 years ago. A small amount of breeding took place with European Neanderthals and Russian Denisovans before they became extinct.

Above: Studies have revealed that we are all descended from a handful of ancestors. For instance, the Y-chromosome of Native American peoples has revealed a single Native American man, from whom 85% of all Native Americans in South America, and half those in North America, are directly descended.

INITIATION AND COOPERATION
right from the start

The origin of the first strand of DNA or RNA on Earth remains a mystery. It may have arrived from elsewhere, but even so, somewhere in the universe, nucleic acids must have been conjured from a primal ooze, perhaps in hydrogel during lightning strikes, or in deep hot fissures, underground or underwater. Lipid bilayers, the basic structure of cell walls, spontaneously form from phospholipids, and one strand of nucleic acid seems to have had the correct coding for manufacturing these. Thus was born the first organism (*below*), which multiplied until its clones and variants began teaming up with or competing against one another. Symbiotic relationships between single-celled organisms like bacteria and archæa (*the two kinds of prokaryotes*) became formalized when colonies of cells began sharing their DNA-strands in a nucleus, giving rise to more complex, multicellular organisms, called eukaryotes (*see pages 224–225 and 228*).

The story of life is, therefore, one of cooperation as much as competition. Cells are specialized for different tasks, like human beings (indeed human culture may be seen as uniquely resulting from specialization, or the division of labour). By trying new kinds of cells, forms, partnerships, and energy sources, DNA hosts managed to leave their watery origin and survive in a wide variety of habitats, always carrying the currently successful versions of the DNA code around with them.

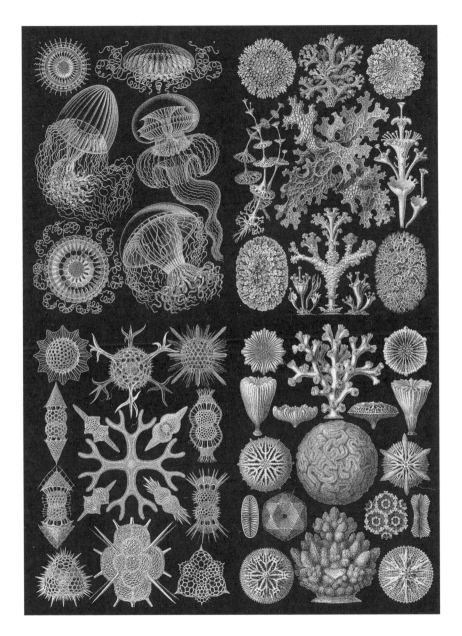

PARASITISM AND SYMBIOSIS
the human question

Many organisms develop relationships with others, either *parasitic* or *symbiotic*. In parasitic relationships only one species benefits from the arrangement, while symbiotic relationships suit both or all parties. Lice, fleas, and worms, for example, afford no advantage to their hosts, but gain resources themselves, while the bacteria in our stomach help us digest, and get fed in the process. Although some deadly bacteria and viruses (*below*) seem to have a lose-lose dynamic with their hosts, they aim to infect new victims before both die.

Some symbiotic relationships are so involved that they result in composite organisms. An example of an animal-animal composite is the Portuguese man-of-war, which looks like a jellyfish. The components of its body are actually different species of organisms working in a cooperative colony. An example of a plant-plant composite is a lichen, which is part alga and part fungus. There are also animal-plant composites, such as upside-down medusæ (*see previous page, top left*), which are jellyfish that contain colonies of algæ.

Mankind may be increasingly likened to a parasite with respect to most life on Earth. Meanwhile, we are in a symbiotic relationship with daffodils, apple trees, dogs, cows, chickens, grasses, and a few other species, all of which thrive at the expense of the many.

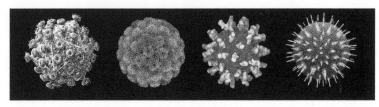

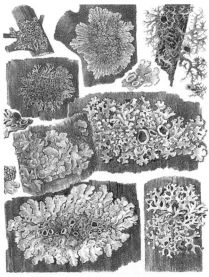

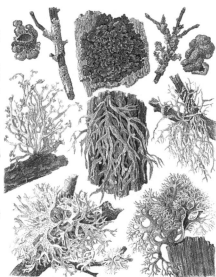

Above, and top right page 211: Various species of lichen. Lichens are composite organisms consisting of a symbiosis between fungi and photosynthetic algæ partners which derive food for the lichen from sunlight.

Left: A Portuguese man-of-war. This is a siphonophore, a colony of specialized polyps and medusoids.

Below: The household flea. An example of a parasite, bringing no gain to its host, simply feeding off it.

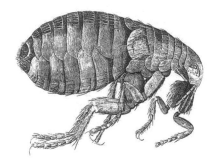

KIN KINDNESS
all for one and one for all

Many creatures perform seemingly selfless acts which protect the genome of their close kin, or even their entire tribe. Altruism operates as a mechanism in evolution because adaptations can also occur at the group level. Although personal selfishness tends to beat altruism within a group, altruistic groups generally beat selfish groups. Altruistic behaviour ensures the survival of group characteristics, both genetic and memetic.

In social insects (ants, wasps, bees, and termites), sterile workers devote their entire lives to the queen, with no chance of breeding themselves. Vampire bats (*opposite top*) share blood with hungry neighbors on returning to their caves, but keep a tally, and expect a return of the favour. Altruism has behavioural mottos: "family first", "help your friends and they'll help you", "safety in numbers", "care for the sick and elderly". A vervet monkey will make an alarm call to warn chums of an unnoticed predator, despite making itself a target.

Extending their sense of kin, dogs occasionally adopt orphaned cats, squirrels, or ducks. Dolphins have been known to assist sick, injured, or struggling animals, their extended evolutionary kin.

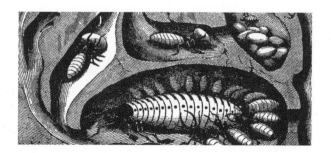

Above: Vampire bats exhibit one of the most universal kinds of altruism, reciprocity. Tending to roost in the same place every night, they get to know each other and will regurgitate blood for a hungry neighbor if the favour is likely to be reciprocated. Vervet monkeys, likewise, will come to the aid of others who have helped them in the past.
Below: Walruses will adopt and rear young calves from different families whose parents have died.

SEXUAL SELECTION
the beautiful things creatures find attractive

Sexual reproduction introduces variation beautifully, which is why most complex species use it. Even non-sexual species occasionally employ some form of sexual reproduction to avoid stagnation. Species that reproduce asexually (without meiosis or mitosis) or by parthenogenesis (without male fertilization) do not produce the genetic variation necessary for effective natural selection.

Males in some animal species hope to mate with as many females as they can, ensuring the wide dispersal of their DNA through their billions of sperm. Meanwhile, a female who has to devote significant energies to the survival of her offspring has fewer opportunities to pass on her DNA. Because she can only have a small number of offspring, it is generally in her interest to look for the best DNA available. As a result she is often choosier than a male, and can require proof of quality, or seduction. This can lead to amazing results.

The classic example is the peacock's tail (*opposite*). It is a totally cumbersome disadvantage to peacocks in every respect apart from its sexiness to peahens. The fittest peacock with the tidiest and most mesmerizing display gets to pass on his genes.

Female guppy fish find colourful males irresistible, especially those with prominent collars. So sexual selection gradually produces populations of colourful fish until the time that predators move in and easily target them. Natural selection leaves the duller ones.

In stag beetles (*opposite*) the outsized and largely useless mandibles are the product of sexual selection by females. They are used by the males to fight over the females, but the males select themselves, the females mating with whomever wins.

CONVERGENT EVOLUTION
inevitable solutions

Variation generally produces evolutionary divergence, as species adapt to fill as many eco-niches as they can. However, it turns out there are often only a limited number of good design solutions when it comes to solving problems such as how best to fly (*below*) or see (*opposite top*), and that is why many species share similar features. When not due to a directly shared ancestry (i.e. coded for by the same DNA) this is described as *convergent* evolution, as analogous features have evolved completely independently.

Convergent evolution can result in both *collective* and *singular* analogous traits. In marsupial and placental mammals, corresponding species that have adapted to fill similar eco-niches on different continents generally resemble one another despite being genetically distant. An example of a singular analogous trait is the camera eye of vertebrates and cephalopods, which is coded for by different DNA yet virtually identical in structure and function in both species (except in the former it grows out of the brain, and in the latter it grows toward it, which is why the octopus doesn't have a blind spot). Groping around for improvements, evolution again and again finds the same solutions that work best.

Left: Three out of the many different camera eyes that have evolved completely separately in different species in different parts of the world (after Conway Morris).

At the top, the human eye focuses by changing the shape of its lens. In the center, the eye of an octopus (a cephalopod mollusc) focuses by moving its lens forward and back. At the bottom, the eye of a marine annelid, a relative of the earthworm. Lens eyes have also evolved independently in the brainless cubozoan jellyfish, dinopis spiders, and heteropod snails, to name but a few examples.

Optic nerve
Retina
Pigmented layer
Nuclear layer

This page, and opposite: Examples of convergent evolution. In the matrix of all possible solutions to problems, only some work, and only some of these work well. Whether it is the discovery of a chemical which responds to light, an efficient method of propulsion, or an outer form maximized for lack of drag through water, the best solutions are limited in number, and constrain the outcomes. The idea is not dissimilar to Plato's ideal forms, perfect solutions, shadows of which appear on Earth.

Above: Marsupial mammals have independently evolved similar forms to non-marsupials. There are marsupial deer (kangaroo), squirrels (koala), rabbits (bandicoots), rats, and mice.

Above: Sharks and crocodiles are perfected forms which have changed only slightly in the last 100 million years. Isolated niche species in Lake Tanganyika have evolved to almost exactly resemble their ocean counterparts.

DEATH
and other helpful illnesses

Death is something everyone can be sure of. However, cells can theoretically reproduce infinitely (some have been kept alive for decades in laboratories), keeping the body young and strong, so why are they programmed to die? Why is it that each cell can only replace itself a certain number of times? It turns out the protective tip of each chromosome (*the telomere, opposite top*) shortens on each copying, like a fuse, and when it is gone, decay and death follow. Death is useful. As we get older our DNA accumulates errors, and death stops these errors being passed on. Sex and death drive evolution. Thus, while women safely create all their ova before they are born, men's sperm become ever more error-ridden as they age. Even long-lived tortoises (150 years) and crocodiles (100 years) occasionally need to be refreshed from their gene pools.

Some illnesses also confer advantage. A famous example is sickle-cell disease, which protects against malaria (*lower opposite*).

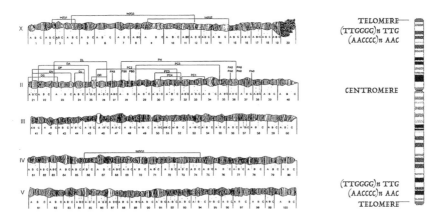

X — A B C A B C A B C A B C A B C A B C A B C A B C A B C A B C A B C A B C
1   2   3   4   5   6   7   8   9   10 11 12 13 14 15 16 17 18 19 20

II — A B C A B C A B C A B C A B C A B C A B C A B C A B C A B C A B C
21 22 23 24 25 26 27 28 29 30 31 32 33 34 35 36 37 38 39 40

III — A B C A B C A B C A B C A B C A B C A B C A B C A B C A B C A B C
41 42 43 44 45 46 47 48 49 50 51 52 53 54 55 56 57 58 59 60

IV — A B C A B C A B C A B C A B C A B C A B C A B C A B C A B C A B C
61 62 63 64 65 66 67 68 69 70 71 72 73 74 75 76 77 78 79 80

V — A B C A B C A B C A B C A B C A B C A B C A B C A B C A B C A B C
81 82 83 84 85 86 87 88 89 90 91 92 93 94 95 96 97 98 99 100

TELOMERE
$(TTGGGG)n$ TTG
$(AACCCC)n$ AAC

CENTROMERE

$(TTGGGG)n$ TTG
$(AACCCC)n$ AAC
TELOMERE

Above: At the tip of every chromosome are regions of repetitive DNA called telomeres that protect the ends. In vertebrates, fungi, and even some slime molds, the repeated motif is TTAGGG. In insects it is TTAGG. The number of repeats varies (around 1,000 in humans). Every time a cell divides, the copying mechanism misses some repeats and shortens the chromosome. When is it all gone, the cell cannot reproduce, only die. Death is part of the program.

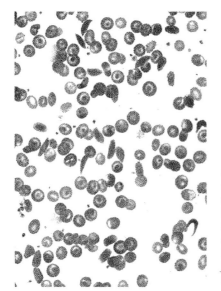

Left: Sickle-cell disease is a genetically inherited condition which reduces life expectancy because the deformed sickle-shaped red blood cells it produces inhibit gaseous exchange in the lungs. The disease reduces the life expectancy of those who have it, so why it has persisted? The answer is that it confers a resistance to malaria, transmitted by mosquitoes (above). So in places where people are more likely to die from malaria than sickle-cell, populations tend to have unusually high instances of the disease.

MIMICRY AND CAMOUFLAGE
the advantages they confer

Many animals use nature's own visual language: scaring away predators by looking more dangerous than they really are, or disappearing into the background, vanishing from hungry eyes.

Camouflage is a form of mimicry where animals evolve to mimic their surroundings to improve their prospects of survival as predator or prey. In animals its effectiveness can rely on both the correct appearance and appropriate behaviour.

Other forms of mimicry involve species mimicking each other for a number of successful reasons. *Batesian* mimicry is where harmless species mimic harmful ones. Wasps, for example, are harmful and have aposematic (warning) black and yellow stripes. A number of moths, beetles, and hoverflies mimic them, resulting in birds steering clear of them for fear of being stung. In *Müllerian* mimicry, species mimic one another for mutual benefit. It is seen in similarly marked tropical butterflies, which are all distasteful to birds. In *Mertensian/Emsleyan* mimicry, deadly prey mimic a less dangerous species. This is because they are so poisonous that predators always die from their bites, never getting a chance to learn to avoid them. Certain deadly coral snakes mimic other snakes that are less harmful (*below*).

Mimicry is occasionally seen in plants too. Some tropical vines have fake butterfly eggs on their leaves, so that female butterflies lay their real eggs elsewhere.

King Snake = harmless

Coral Snake = deadly

Arizona Coral Snake = poisonous

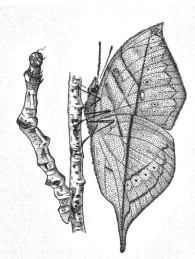

Above: Batesian and Müllerian mimicry. The harmless hoverfly finds it useful to have the warning markings of a stinging tree wasp—an example of Batesian mimicry. The potter wasp displays warning stripes, not dissimilar to those of tree wasps, an example of Mullerian mimicry.

Above: Clever camouflage. On the left, the moth larva ennomos has evolved so it can resemble a twig. On the right, the Indo-Malayan butterfly kallima which has evolved leaf-like markings that help it hide from predators.

Above, and camouflaged left: The sargassum weed-fish has evolved extraordinary markings, protrusions, and appendages, all of which make it extremely difficult for its predators to spot when it hides amongst the floating sargassum weed in the open zone of the Sargasso Sea.

THE KINGDOMS
serial endosymbioses

The first beings, formed from the organic brew of the nascent Earth, inhabited a world of brimstone, hot mud, and marsh gas. These archæbacteria, primarily creative photosynthesizers and scavenging methane producers, together stabilized and slowly began to regulate the planetary environment.

For over a billion years there was an ample oceanic mineral sink for all the oxygen produced, until incendiary levels extinguished most life, driving the old anærobes into specialist underworld niches. Lifeforms which thrived on oxygen diversified into new breeds of mobile consumers, some of whom incorporated other prokaryotes, to mutual benefit. From here on, a succession of such endosymbiotic relationships generated a panoply of protoctists, and different streams of these, with characteristic capabilities, are the ancestors of the three multicellular kingdoms; Fungi, Animalia & Plantæ (*process shown in the diagram opposite, adapted from Lynn Margulis*).

Thus, all complex organisms are vast colonies of ancient bacteria in refined, specialized, and organized relationships. Some of the original forms are now organelles within modern eukaryotic cells, but many exist in looser relationships, such as the microflora of any healthy digestive system. Anærobic methanogens ferment grasses in the stomachs of ruminants. Similar microbes living within protoctists in the guts of termites enable them to digest wood and perform wondrous feats of demolition and architecture, while the displays of fireflies and cuttlefish, medusæ, and angler fish, all depend on an enzyme luciferase, made by archaic sulfur bacteria living within all bioluminescent organisms.

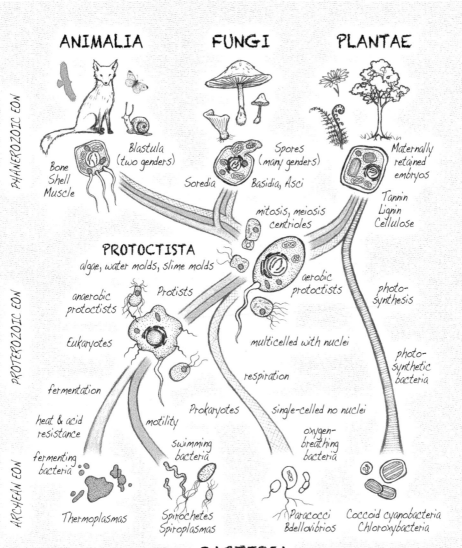

ANIMALIA   FUNGI   PLANTAE

PHANEROZOIC EON

Bone
Shell
Muscle

Blastula
(two genders)

Spores
(many genders)

Soredia

Basidia, Asci

Maternally
retained
embryos

Tannin
Lignin
Cellulose

mitosis, meiosis
centrioles

PROTOCTISTA
algae, water molds, slime molds

Protists

anaerobic
protoctists

aerobic
protoctists

photo-
synthesis

PROTEROZOIC EON

Eukaryotes

multicelled with nuclei

photo-
synthetic
bacteria

fermentation

respiration

heat & acid
resistance

motility

Prokaryotes

single-celled no nuclei

fermenting
bacteria

swimming
bacteria

oxygen-
breathing
bacteria

ARCHEAN EON

Thermoplasmas

Spirochetes
Spiroplasmas

Paracocci
Bdellovibrios

Coccoid cyanobacteria
Chloroxybacteria

BACTERIA

PROKARYOTES
and the tree of life

Unlike eukaryotes (multicelled beings with nuclei), single-celled prokaryotes carry their genes in nucleic acid toruses unconstrained by any envelope and readily share this information with each other, or scavenge it wherever they find it. This cooperative genetic promiscuity and disregard for individual identity, coupled with the rapid reproduction of their simple wee forms, enables them to adapt flexibly and robustly to severe environmental changes. They have been here for around 4 billion years, adapting easily to changes, so attempts at eradicating bacteria merely accelerate their ongoing evolution and enhance their resilience.

Whilst they tend to one of three shapes: rods (bacilli), balls (cocci), or twists (spirochætes), prokaryotic mutability defies any finite system of classification, and bacteria instead constitute a vast, global community throughout the whole of life. They have flourished in every habitat, mastered all essential biochemical processes, and are fundamental to the planet. Bacterial fixation of atmospheric nitrogen provides for all the matter for proteins, genes, and ultimately hormones and natural medicines. Ærobic prokaryotes adapted the potential of oxygen originally liberated by cyanobacteria, yet anærobic archæbacteria still prosper in obscure realms akin to the early Earth; harnessing volcanic sulfur vents, making methane in swamps, and evaporite salt mats.

With the evolution of larger lifeforms, suitable habitats formed inside eukaryotes, and indeed wherever they have arisen, prokaryotes have diversified to colonize them. The evolution of these communities remains dynamically shaped through their own relationships and those formed with eukaryote sentinel systems, characterized in health by long-term familiarity, beneficial mutual respect, and tolerance.

THE
TREE OF
LIFE

EUKARYOTES

PROKARYOTES

?

Above: The Tree of Life, showing the main branches; Bacteria, Fungi, Plants, and Animals. Even more simple than prokaryotes are viruses and prions, which are regarded as non-living, despite their status as organic entities. Viruses are bundles of nucleic acid within a shell and cannot grow or reproduce outside of host cells. Most parasitize the cells of eukaryotic organisms. Retroviruses transfer their DNA into the chromosomes of their hosts, others viruses invade bacteria and are known as bacteriophages (bacteria-eaters). Prions lack nucleic acid and shells; being little more than particles of protein, they duplicate themselves either inside or outside the cells of host organisms.

PROTISTS
free expression

Protoctista is the fundamental eukaryotic kingdom; a panoply of aquatic, microscopic unicellular Protists, including ciliates, mastigates, amœbæ, algæ, and diatoms, plus their larger scale colonies, such as slime molds and kelps. Amidst the chemical communications, casual genetic exchanges and entire consumptions of one prokaryote by another, eukaryotes evolved through serendipitous permutations of resources combined within a single cell. Ancestrally they were underworld anærobes, an archaic thermoacidophile, contributing the heat and acid tolerant membranes, nucleic acid, and proteins for a fermenting iron and sulfur metabolism, incorporating a swimming spirochæte, with a cytoskeletal infrastructure of dynamically organizing microtubules, thus an internal system of locomotion and transport, with surface cilia adding motility and local control.

Over time, organelles slowly enhanced their cellular complexity, compartmentalizing chaotic processes in stable spaces to transform, transport, and conserve resources, as genetic material was now isolated inside a nuclear membrane with a framework for operation and division. The cellular inclusion of an ærobically-respiring purple bacterium around 2 billion years ago resulted in mitochondria and the capacity to tolerate and utilise oxygen, so new protists then quickly proliferated into any watery niches with some organic matter. Among this diversity were ancestors of animals and fungi. Plant ancestors, meanwhile, required photosynthetic green bacteria to evolve into chloroplasts within free-swimming green algæ, achieving energetic autonomy for eukaryotes now able to seek and use light energy to bind low order CO_2 into their own higher order macromolecules.

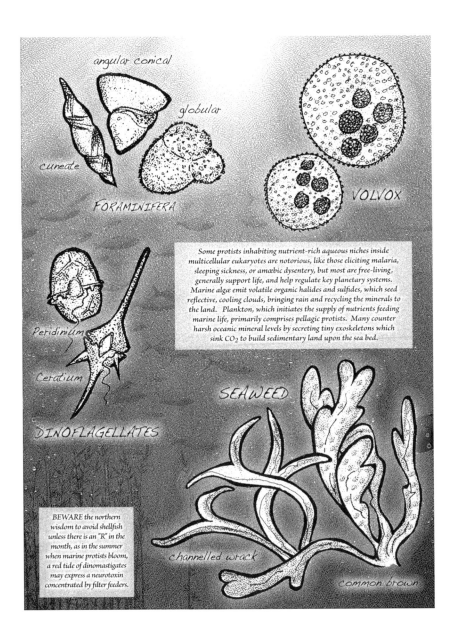

angular conical

globular

cuneate

FORAMINIFERA

VOLVOX

Peridinium

Ceratium

DINOFLAGELLATES

Some protists inhabiting nutrient-rich aqueous niches inside
multicellular eukaryotes are notorious, like those eliciting malaria,
sleeping sickness, or amœbic dysentery, but most are free-living,
generally support life, and help regulate key planetary systems.
Marine algæ emit volatile organic halides and sulfides, which seed
reflective, cooling clouds, bringing rain and recycling the minerals to
the land. Plankton, which initiates the supply of nutrients feeding
marine life, primarily comprises pellagic protists. Many counter
harsh oceanic mineral levels by secreting tiny exoskeletons which
sink CO_2 to build sedimentary land upon the sea bed.

SEAWEED

BEWARE the northern
wisdom to avoid shellfish
unless there is an "R" in the
month, as in the summer
when marine protists bloom,
a red tide of dinomastigates
may express a neurotoxin
concentrated by filter feeders.

channelled wrack

common brown

FUNGI
one man's waste is another man's fertilizer

Fungi are supreme recyclers. Although as lichens, in strict symbiotic partnerships with cyanobacteria or algæ, they may reap the fruits of photosynthesis directly, most form a filamentous microscopic network throughout the soil of the planet where, in communities with underworld bacteria, they excel at transforming organic matter, mining minerals, and divining water. Inside a wall of cartilagenous chitin, the same material as insect exoskeletons, the individual cell membranes within the fungal mycelium fuse as a continuum, producing a single, multinucleate protoplasmic animal in a tube, which streams outward from branching hyphal tips leaving its wake devoid of nutrients. Hence the fruiting bodies, which indicate subterranean fungal sex, manifest macroscopically as mushrooms in fairy rings.

In high sugar concentrations, including those inside multicellular eukaryotes, fungi typically adopt a unicellular yeast form in suspension, and many, as facultative anærobes, can switch respiratory mode to exploit oxygen if present. For example, with oxygen Saccharomyces cerevisiæ can respire sugars completely to gaseous CO_2, thereby leavening bread, but obtains energy in its absence by fermenting them to ethanol.

Adept at amino acid chemistry, many fungi convert tyrosine to tyramine, a stimulant concentrated in smelly cheeses, pickled and gamey meats and fish, live alcoholic drinks, and yeast extracts. Claviceps purpurea, ergot or rust fungus, grows on grain, especially well under moist conditions, and produces diverse indole alkaloids from tryptophan, including psychedelic lysergides, powerful vasoconstrictors, and uterine stimulants, littering history with episodes of St Antony's Fire and witch trials, but still treating migraines and controlling bleeding at childbirth.

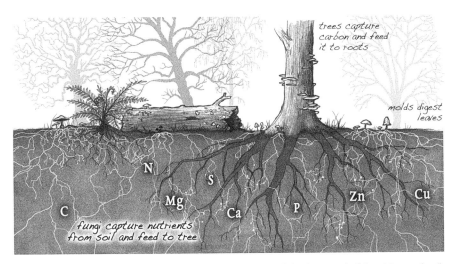

trees capture
carbon and feed
it to roots

molds digest
leaves

N

S

Mg

Ca

C

P

Zn

Cu

fungi capture nutrients
from soil and feed to tree

Above: Soil fungi readily decompose complex proteins and carbohydrates, including keratin and cellulose. Many can handle plant polyphenolics, as in wood, but most struggle with conifer resins. Particularly fond of sugars, they often form symbiotic mycorrhizal relationships with plants, which receive water, minerals, and soluble nitrogen, enabling them to thrive, along with auxins, plant hormones which direct sugars from manufacture in the leaves toward the roots and reward the fungus.

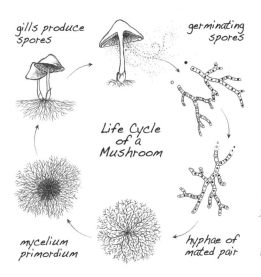

gills produce
spores

germinating
spores

Life Cycle
of a
Mushroom

mycelium
primordium

hyphae of
mated pair

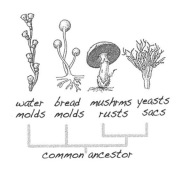

water   bread   mushrms   yeasts
molds   molds   rusts   sacs

common ancestor

Fungi reproduce through robust, resilient spores produced asexually to survive adversity, or sexually, when conditions favour. Distribution typically uses the elements, but many have strategies to ensure growth from a suitably moist, nutritious environment, like attaching spores with mucus to grass tips to be deposited a couple of days later in a pile of herbivore manure.

PLANTS
non-vascular and vascular

The plant and animal kingdoms exist in dynamic equilibrium, relying on one another for their survival. Both use DNA (*below*) in the same way. Plants photosynthesize carbon dioxide and water into oxygen and glucose in sunlight, at the same time respiring, the opposite process, like animals do exclusively. Plants provide food at the base of the food chain for animals, while animals return nutrients to the soil by way of excreta and decomposition. Plants also bring new stocks of nutrients into the equation by deriving them from minerals, water, and air. This interdependency between organisms is a vital characteristic of life on Earth, which is why the planet's ecosystem should be respected.

Plantæ, the plant kingdom, is divided primarily between vascular and non-vascular plants. Vascular plants have the ability to transport fluids from roots to other parts of the organism, which means they can live on land and colonize areas where there is no surface water. Vascular plants include clubmosses, ferns, horsetails, conifers, cycads, ginkgos, and flowering grasses, rushes, herbs, shrubs, and all trees. There is a clear evolutionary progression from the production of spores to naked seeds and then to shelled seeds and nuts, which provide the germinating plant embryos with food and protection. Non-vascular plants are the most simple of land-dwelling plants; generally small and shade-loving, they have no roots, stems, or leaves.

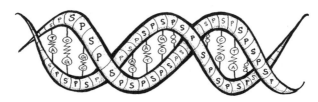

233

PLANT MORPHOLOGY
roots in the earth, leaves in the sky

The ciliated protists from which plants evolved propelled themselves between the aquatic sediment, where they acquired minerals, and the surface, where they absorbed sunlight. For terrestrial plants, accessing these resources from a single site is the primary concern. As embryonic plants grow from primordial meristem, some cells push upward toward the light, some move with gravity underground, and between the two cambium cells differentiate into vascular xylem and phloem, connecting the extremities, with storage cells at the core. From this initial set up, modules of root, stem, and leaf branch at characteristic frequencies, with variations such as tactile tendrils, protective thorns, and flower buds.

The survival of the whole is not invested in any one module, allowing for generous donations of tissue, without fatality, to passing herbivores or parasites, and as long as some meristem persists, the whole plant can regenerate. This enables plants to reproduce asexually from runners, rhizomes, stolons, and suckers, and permits pruning and harvesting.

Stems elevate leaves into position to optimize solar exposure, presenting flowers to their vehicles of pollination and seeds to theirs of dispersal. Many are supported by calcium minerals or woody lignin, with a protective surface layer of bark which commonly houses whole communities. In a parabolic pattern symbolizing the path of the Sun, individually rotating leaves combine in a mosaic presenting a shape which continually maximizes the overall photosynthetic surface. Flowers are evolved for plant sex, wherein fertilized ovules develop into seeds within fruit derived from the ovary, yet often communicate with the animal kingdom, synthesizing resonant and seductive pigments, scents and nectar, which diversity underlies their Linnaean classification.

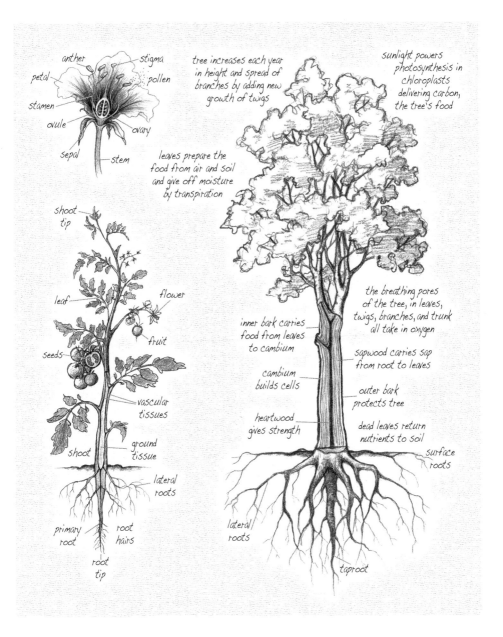

anther

stigma

petal

pollen

stamen

ovule

ovary

sepal

stem

tree increases each year in height and spread of branches by adding new growth of twigs

sunlight powers photosynthesis in chloroplasts delivering carbon, the tree's food

leaves prepare the food from air and soil and give off moisture by transpiration

shoot tip

leaf

flower

fruit

seeds

vascular tissues

shoot

ground tissue

lateral roots

primary root

root hairs

root tip

inner bark carries food from leaves to cambium

cambium builds cells

heartwood gives strength

the breathing pores of the tree, in leaves, twigs, branches, and trunk all take in oxygen

sapwood carries sap from root to leaves

outer bark protects tree

dead leaves return nutrients to soil

surface roots

lateral roots

taproot

INSIDE PLANTS

behold the colours

Plants harness solar energy to make the complex organic compounds which feed and direct all eukaryotic life. This photosynthesis begins in stacks of membranes called thylakoids in chloroplasts (*lower opposite*), where complexes of light-harvesting pigments undergo sequential photon excitations and emissions to yield quanta perfect for splitting water. Chlorophyll, the complex of pigments in chloroplasts, absorbs red and blue light, so appears green. Plants also use red, orange, and yellow carotenoids to extend the range of usable light frequencies, and these lipids are used in your retina as derivatives of vitamin A from the plants you eat, resonating with photons to define colour vision.

The light energy captured by chloroplasts is used to separate protons from electrons across the thylakoid membrane. Oxygen is the final destination of these electrons, and is expired as O_2, while discharging the proton gradient regenerates ATP (*see page 393*), used to power all sorts of biochemical transformations. The carbon liberated from CO_2 is fixed into skeletons of small sugars like glucose and fructose, which are combined into larger structural sugars, like cellulose, or stored as energy reserves in the amyloplast starches of roots, tubers, and seeds.

At night, plants make a huge range of phytochemicals using the energy harnessed during the day, transforming the small sugars into amino acids with help from soluble nitrogen from the roots, and thus into medicinal or toxic alkaloids, related phenolics like vanilloids, salicylates, and balsams, and polymers like astringent tannins and woody lignin. Plants can also reduce sugars to lipids to make the edible oils of seed energy stores, protective soaps, and aromatic oils like menthol, limonene, and camphor, and resins and latexes like myrrh and rubber.

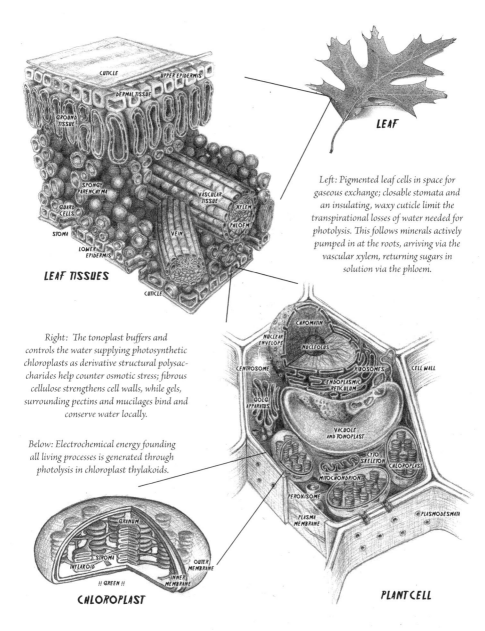

CUTICLE

UPPER EPIDERMIS

DERMAL TISSUE

GROUND
TISSUE

SPONGY
PARENCHYMA

VASCULAR
TISSUE

XYLEM

GUARD
CELLS

PHLOEM

STOMA

VEIN

LOWER
EPIDERMIS

LEAF TISSUES

CUTICLE

LEAF

Left: Pigmented leaf cells in space for gaseous exchange; closable stomata and an insulating, waxy cuticle limit the transpirational losses of water needed for photolysis. This follows minerals actively pumped in at the roots, arriving via the vascular xylem, returning sugars in solution via the phloem.

Right: The tonoplast buffers and controls the water supplying photosynthetic chloroplasts as derivative structural polysaccharides help counter osmotic stress; fibrous cellulose strengthens cell walls, while gels, surrounding pectins and mucilages bind and conserve water locally.

Below: Electrochemical energy founding all living processes is generated through photolysis in chloroplast thylakoids.

CHROMATIN

NUCLEAR
ENVELOPE

NUCLEOLUS

CENTROSOME

RIBOSOMES

CELL WALL

ENDOPLASMIC
RETICULUM

GOLGI
APPARATUS

VACUOLE
AND TONOPLAST

CYTO
SKELETON

CHLOROPLAST

MITOCHONDRION

PEROXISOME

PLASMODESMATA

PLASMA
MEMBRANE

GRANUM

STROMA

OUTER
MEMBRANE

THYLAKOID

!! GREEN !!

INNER
MEMBRANE

CHLOROPLAST

PLANT CELL

POLLINATION AND DISPERSAL
manipulating the animals

To benefit from sexual reproduction, plants need ways to ensure male gametes (pollen) are delivered to female gametes (ovules) in other plants of the same species. The resulting zygote then has to be carried off to a distant spot to grow. Initially, the four elements were the vectors of dispersal; wind wafting lightweight pollen to other plants or assisting the helicopter flight of sycamore keys, and water dispersing seeds as large as coconuts to travel the oceans and colonize the globe. With the arrival of animals on the scene, opportunities for long-distance travel rocketed, and flowers developed ingenious ways to manipulate insects, other arthropods, birds, and humans to these ends.

The reward for the pollinator is sometimes a small meal of nectar and pollen, but often nothing. For instance, some orchids fool spiders into trying to have sex with them, while flowers like Amorphophallus release volatile diamines like putrescine and cadaverine to attract flies who think they are landing on some rotting meat. Nectar is generally difficult to reach, to ensure the bird or bee delves into the flower and gets covered in or delivers pollen to the flower, whose colour, scent, taste, and sometimes UV landing strips all serve to attract mobile animals.

Seed dispersal can be as simple as a hooked burdock ball that attaches to animal fur, or refined single species relationships, as smelly durians have with their distributors, bats. Humans have selected the most generous fruit producers and the prettiest flowers, but have also been manipulated by plants like cannabis, coca and poppies which make use of human biochemistry to propagate themselves. Some plants even start bushfires with substances like ethene to clear the way for their young.

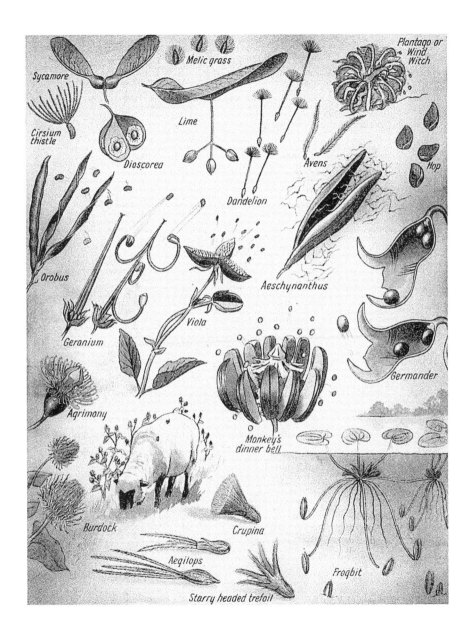

Sycamore

Melic grass

Plantago or Wind Witch

Cirsium thistle

Lime

Dioscorea

Avens

Hop

Dandelion

Orobus

Aeschynanthus

Viola

Geranium

Germander

Agrimony

Monkey's dinner bell

Burdock

Crupina

Froqbit

Aegilops

Starry headed trefoil

ANIMALS
mostly unknown insects

Animals range from simple unicellular organisms to highly complex multicellular organisms. They are motile, able to move spontaneously and independently (at least at some point in their lives) and they, like plants, consist either of single cells or collections of cells that communicate and cooperate with one another through various complex signalling pathways. Most animal phyla appeared in the seas of the Cambrian era, around 550 million years ago (*see page 395*).

The classical taxonomic division of the kingdom Animalia is into vertebrates and invertebrates (with or without spines). The invertebrate group comprises about 97% of all animal species and includes amoebas, hydras, sponges, worms, molluscs (slugs, snails), cnidarians (jellyfish, anemones, corals), echinoderms (urchins, starfish), cephalopods (squid, octopus, cuttlefish), and numerous arthropods (crustaceans, arachnids, insects). The vertebrate group includes fish, amphibians, reptiles, birds, marsupial mammals, and placental mammals. Each extant species is at the apex of its own evolutionary story. Others didn't make it this far.

The modern classification of the Animal kingdom involves 13 phyla, including at least three containing different kinds of worms. The largest phylum by far is Arthropoda, mostly populated by insects, with well over a million named species, and 20 million unnamed. In all, there are probably around 30 million species of plants and animals on Earth, of whom human activity is killing off about 50,000 a year, or 1% every 6 years, the fastest rate of genomicide since the K-T extinction which wiped out the dinosaurs along with 85% of all species on Earth. Last time it took 30 million years for the Earth to recover.

241

WHAT IS AN ANIMAL?
a moving eating archipelago

Animals are predators of organic matter—they survive by eating other living beings or their products, and rely on the previous hard work and ingenuity of their fodder to obtain the nutrients they need. They breathe oxygen for their mitochondria to burn some of this food, which provides the energy to drive all the older anaerobic biochemical processes their cells have incorporated. A small free-floating opportunistic nibbler leads an uncertain life, so animals need reliable ways to encounter and ingest food. The most ancient animals like sponges, corals, and jellyfish do this by using their cilia to waft seawater through themselves, from which they feed (mainly on bacteria and plankton) and extract oxygen, the seawater also washing away their leftovers. Most corals house endosymbiotic algæ inside their cells to help with all this.

All animals exhibit radial or bilateral symmetry, except sponges which exhibit none. The evolution of forms able to move around looking for things to kill or graze upon took the transition from the two-layer radial symmetry of jellies and anemones to the three-layer bilateral symmetry first seen in the trilobites of the Cambrian era when compact, segmented forms with a head and tail, and so a mouth and anus, and front and back, first emerged. Such a shape favours streamlining and motility, and cephalization, with sensory and control nerves concentrated at the head, which eventually means a brain. Animals followed plants onto the land over 400 million years ago in order to eat them, so had to enclose their own private mini-sea, modified and maintained, to bathe their cells in, and develop the apparatus to exchange gases with air rather than water. The earliest animal known to have hobbled ashore was a proto-millipede, near what is now Stonehaven in Scotland.

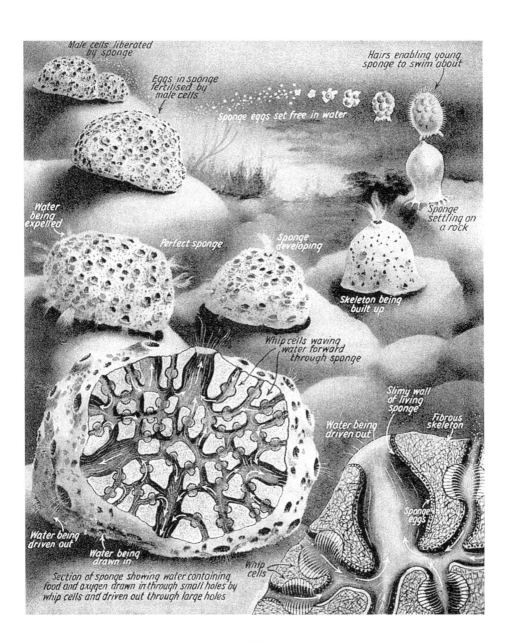

Male cells liberated by sponge

Hairs enabling young sponge to swim about

Eggs in sponge fertilised by male cells

Sponge eggs set free in water

Sponge settling on a rock

Water being expelled

Perfect sponge

Sponge developing

Skeleton being built up

Whip cells waving water forward through sponge

Slimy wall of living sponge

Fibrous skeleton

Water being driven out

Sponge eggs

Water being driven out

Water being drawn in

Section of sponge showing water containing food and oxygen drawn in through small holes by whip cells and driven out through large holes

Whip cells

ANIMAL MORPHOLOGY
interiorization, segmentation, skeletalization

Hungry animals want to keep their food to themselves, and digest it fully, so the primary organ to interiorize is the gut, surrounded by the apparatus to serve it and move it around. The development of animal embryos roughly mirrors the march of evolution—from a ball of cells that flattens, becomes layered and segmented, and differentiates as it grows. The sequence progresses to annelids (ringed worms) and onto sea squirts, fishes, amphibians, reptiles, then birds or mammals, stopping or diverging at any stage. Growing in repetitive modular segments permits a continuous but variable assemblage, with paired appendages added on as needed. Genetic evidence points to a common origin for this process; your body is segmented like the rings of an earthworm.

Animal cells have non-rigid membranes, and can grow into dense, infolded, specialized tissues, but need a firmer structure to hang it all on. Earthworms use a simple hydrostatic skeleton, a fluid-filled cavity between the gut and outer muscular tube, which squeezes segmentally to move. Most animals, like arthropods (insects, spiders, crabs) which comprise the vast majority of animal species, have an exoskeleton made of a protein called chitin, both structural and protective, which they have to moult to allow growth. Crustacean cuticles are also mineralized, and serve as calcium dumps, like vertebrate spines and mollusc shells, while sharks and newborn babies have skeletons made entirely of cartilage.

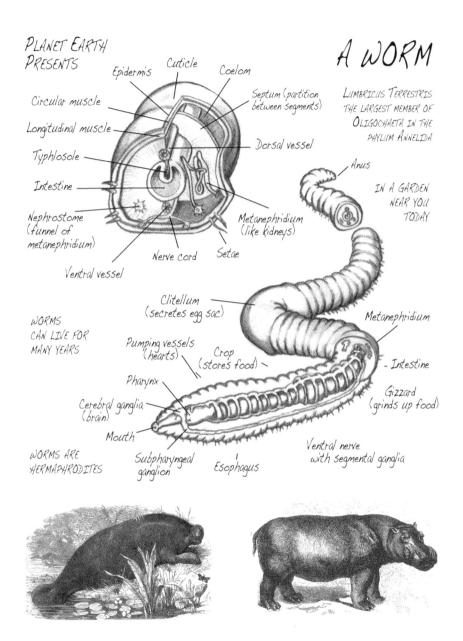

PLANET EARTH
PRESENTS

A WORM

LUMBRICUS TERRESTRIS
THE LARGEST MEMBER OF
OLIGOCHAETA IN THE
PHYLUM ANNELIDA

Epidermis
Cuticle
Coelom

Circular muscle

Septum (partition
between segments)

Longitudinal muscle

Dorsal vessel

Anus

Typhlosole

IN A GARDEN
NEAR YOU
TODAY

Intestine

Nephrostome
(funnel of
metanephridium)

Metanephridium
(like kidneys)

Ventral vessel

Nerve cord

Setae

WORMS
CAN LIVE FOR
MANY YEARS

Clitellum
(secretes egg sac)

Metanephridium

Pumping vessels
(hearts)

Crop
(stores food)

- Intestine

Pharynx

Gizzard
(grinds up food)

Cerebral ganglia
(brain)

Mouth

WORMS ARE
HERMAPHRODITES

Subpharyngeal
ganglion

Esophagus

Ventral nerve
with segmental ganglia

VITAL ORGANS
specialized bits of beasties

Organs began as clusters of specialized cells in flatworms of the warm shallow seas of the Edicaran period, evolving into peripheral gut support systems as true intestines developed. The primæval needs were for circulation, respiration, motion, internal regulation, overall control, and with the evolution of sex, reproduction. The dictates of structure that fashioned the earliest animals in the sea apply equally to the interiorized organs of advanced animals, which inherit their mechanisms of formation. Efficiency of transport, and maximization of surface area and points of contact within a finite space led to certain common shapes—the radiation of nerves and blood vessels resemble a tree; turn a sponge inside out and it's a lung. Similar-looking organs thus may be the result of convergent solutions (*see page 218*).

One consequence of bilateral symmetry is that animals mainly have paired organs, which sometimes fuse, or one may be lost at some stage of growth. Most female scarab beetles, for instance, have only one ovary, but develop two during pupation, one atrophying before the completion of the metamorphosed imago.

MOSS SEX ORGANS
biflagellate moss sperm swim toward the egg like a protist

COBRA VENOM GLAND
its neurotoxin works like bamboo curare and can fell an elephant

VIPERFISH PHOTOPHORE
flashes its bacteria to lure victims and communicate with mates

BUTTERFLY HEART
tubular pumps along dorsal
vessel circulate yellow lymph

DINOSAUR GIZZARD
contained food-grinding stones
like in crocodiles, seals, and birds

GRASSHOPPER OVIPOSITOR
stubby with reinforcing cerci to
burrow a hole to lay the egg in

SHARK TEETH
cut flesh nicely; some sharks
make thousands in one lifetime

ELECTRIC EEL BATTERY
akin to thousands of nerve cell
membranes in a 500 volt stack

EAGLE EGGMAKER
encloses fertilized (or not) ovum
in a food-filled chalk shell

CROCODILE EAR
inner ear is a single intracranial
canal that works amphibiously

FIREFLY GLOWER
flash mimicry by other beetles
lures unlucky males to death

SLIME MOLD GREX
dispersed protists occasionally get
together and act like an animal

SCORPION STING
unharmed by their own venom;
kill thousands of humans a year

MANTIS COMPOUND EYE
transmits hexellated imagery,
forward-facing for hunting

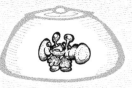

FLEA BRAIN
more a set of ganglions, deciding
mainly when to bite or jump

PRODUCTION
who eats who

Apart from a very few organisms who draw energy from hot oceanic vents, all life on Earth depends either directly or indirectly on energy captured from the Sun. At the bottom of the food chain are producers, or *autotrophs*, plants which convert carbon dioxide and water into glucose through photosynthesis (*page 135*) in their chloroplasts driven by the red and blue frequencies of sunlight. Above them are consumers, or *heterotrophs*, which derive their energy from eating autotrophs or other heterotrophs. Food chains are therefore often organized in trophic levels (*below and opposite top left*). Some consumers, like bacteria and fungi, are *decomposers*, getting their energy from dead organisms which may have been previously chewed up by *detrivores* like snails or vultures.

All living organisms also *respire*, animals breathing out water and carbon dioxide, the opposite to photosynthesis, to maintain essential processes. The *net productivity* of a system is its gross productivity less its respiratory loss (*e.g. leaf example opposite lower left, after Rutherford*).

Apart from the cycling of carbon in photosynthesis and respiration, other essential biogeochemical cycles involve water (*page 357*), nitrogen (*page 145*), sulfur (in proteins and enzymes), phosphorus (in DNA), and other trace minerals. These, through weathering, sedimentation, and uptake by plants or animals licking rocks, are also are passed through the trophic levels as organisms devour and decompose one another.

 1    2    3

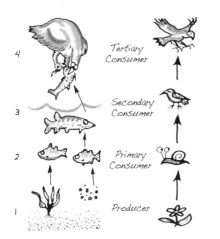

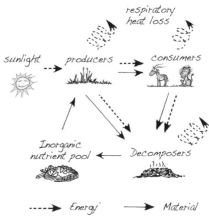

Above: Four trophic levels. Producers capture energy from the Sun. Primary consumers are vegetarians, eaten by secondary consumers. Only 10% of the captured energy is transmitted to the next level up, meaning carnivorous tertiary consumers are relatively rare.

Above: Energy and material flow through an ecosystem. Note how important decomposers are, as they help to recycle nutrients back to primary producers. Soil respiration (caused by bacteria in soil) is an important factor in the global carbon cycle.

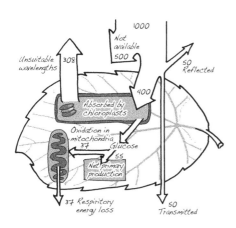

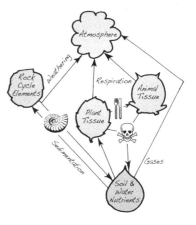

Above: The productivity of a plant. Chloroplasts make glucose from red and blue light (reflecting the green), while mitochondria burn glucose to power cells, causing respiration. From an input of 1,000 units of energy, the gross primary productivity is 92 (37 + 55).

Above: Biogeochemical cycles in the global ecosystem. Around 40 elements cycle through ecosystems, and are vital in supporting life processes. The major reservoir for these slowly moving inorganic materials is normally outside the food chain, in rocks and soil.

Ecosystems and Biomes
the web of life

The web of life on Earth is a single supercomplex ecosystem made of smaller localized ecosystems. An ecosystem is defined as an environment which consists of all of the organisms in an area as well as the inorganic features of that area, including the niches occupied by the various organisms. In any given ecosystem there will exist a food web (*e.g. opposite top right*), normally driven by sunlight and photosynthesis.

Aquatic ecosystems cover 72% of the Earth's surface, and aquatic phytoplankton generate 45% of the world's net primary production, so nearly half of the oxygen in the atmosphere. Aquatic ecosystems are divided into 99% saltwater marine varieties, e.g. oceans, salt marshes, mangroves, and coral reefs, and 1% freshwater varieties, e.g. bogs, swamps, lakes, rivers, ponds (*see pond ecosystem development below*).

Ecosystems can be grouped in *biomes*, climatically similar environments which tend to evolve similar kinds of ecosystems. Based on annual rainfall and temperature, a graph may be drawn of the major land biomes, whether tropical or temperate forests, deserts, taiga, or tundra (*opposite lower right*). Soil development follows biomes and plays a major part in determining the productivity of an ecosystem. A mix of clay, sand, and silt, sometimes with organic matter and microorganisms (fungi and bacteria), soils are fixed in place by plant roots. The perfect mix for optimum productivity is a loam (*opposite lower left*).

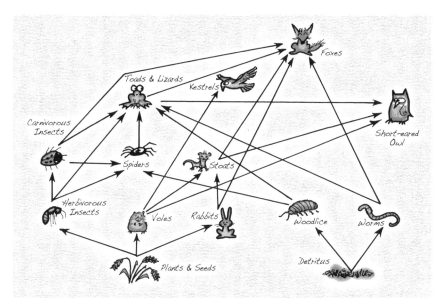

Above: A greatly simplified food web from acid heathland in Dorset, southern England (after Rutherford). Individual insect species are not shown, nor are individual plants or decomposers, but an overall picture of the web may be gained.

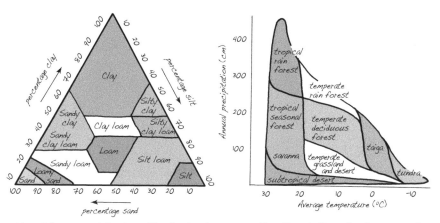

Above: Soils are composed of a mixture of clay, silt and sand, the relative percentages of each of the three giving rise to the various named soil types. Loam is the most productive.

Above: The major biomes plotted against annual precipitation and temperature. Similar types of ecosystems may be found in similar biomes across the world.

THE GAIA THEORY
a living, self-regulating system

The Earth's entire biosphere is today understood as a network of self-regulating systems which maintain the conditions for life on the planet. The salinity in the oceans has been maintained at about 3.4% for a very long time despite rivers adding more salt to the seas. The atmospheric composition has also remained relatively constant for the last 3 billion years, as has the global surface temperature, despite energy from the Sun increasing by 25%. Carbon-based life interacts with and affects Earth's atmosphere using negative feedback loops to maintain homeostasis; e.g. warmer oceans seed more clouds, resulting in cooling (*opposite top*).

Sadly however, idiotic human burning of fossil fuels and decimation of forests for cities and crop fields now seriously threatens the entire system for the first time since the dinosaurs' extinction (*lower opposite*).

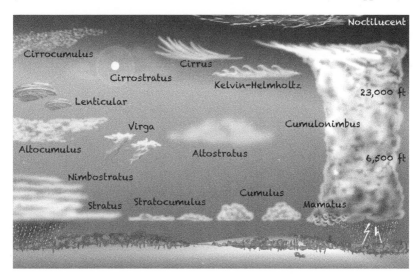

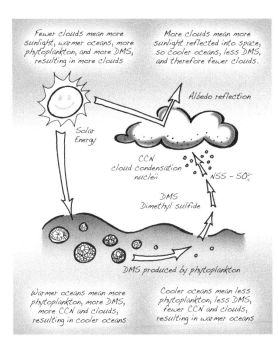

Left: An example of a self-regulating system in Earth's biosphere. The production of dimethyl sulfide by phytoplankton regulates the number of cloud condensation nuclei and so the amount of cloud cover. More sunlight reaching the oceans results in more phytoplankton, more clouds, and so less sunlight reaching the surface of the earth, and vice versa.

Below: The Carbon Cycle (figures show total stored carbon and annual movements of carbon in metric tons). Note the very large part played by soil respiration, more than plants, also the slow absorption of carbon on the sea floor, and then how the emissions from fossil fuels and changing land use are annually increasing atmospheric CO_2. If the global temperature rises much higher then vast quantities of methane stored in the permafrost and under the oceans will also be released, probably leading to runaway global warming.

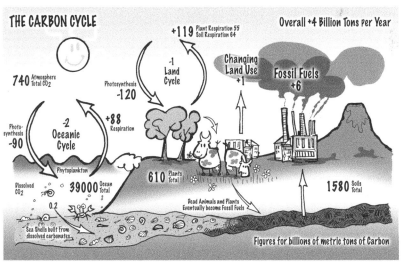

253

ACCELERATED EVOLUTION
genetic engineering and self-evolved code

Humans look like Gaia's best bet for colonizing the galaxy so far, but it might not work out. Memetic agents could eventually select for brawn against brains, or we could wipe out the ecosystems which sustain us. An asteroid or nearby supernova could send us back to the slime age. The eventual colonizer of our galaxy could evolve from some other source than apes (*dolphins, for example, opposite top*).

But then again, genetic engineering (*lower opposite*) could accelerate evolution in the near future, creating enhanced lifespans, features, and new species which could outperform anything we can imagine now.

Evolutionary theory is today used highly effectively in the computer sciences. Dynamic programs spawn randomly varied children that are then selected for target behaviours. In this way robots effortlessly learn to walk in style, fly gracefully, or crawl and wriggle fast (*below*), using *evolved algorithms* which no human has programmed.

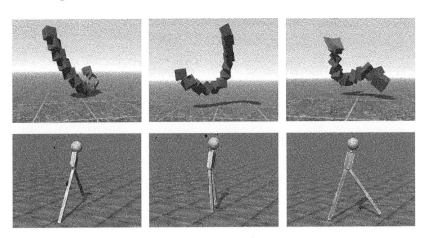

Man is still evolving and the future could take many directions. From ancestral ape to human may only be part of the story. Newly evolved or genetically engineered features may, in the future, create new species. Then again, mankind may be an evolutionary dead end, and it could be another species which later emerges from the global gene pool to colonize the galaxy.

Beans

Ferritin
iron protein

Aspergillus
fungus

Phytase
enzyme

Wild rice

Metallothionin
protein

Daffodil

4 enzymes

increases iron
content

destroys phytate
to allow iron
absorption

adds sulfur
needed for
iron uptake

adds beta-
carotene to form
vitamin A

Above: Transgenic rice, engineered to address the serious iron and vitamin A deficiencies in developing countries.

MEMES
self-replicating thoughts and cultural viruses

Using biological evolution to help better understand cultural evolution, Richard Dawkins, in 1976, invented the concept of the *meme* as the cultural equivalent to the gene, defining it as unit of cultural information existing within a *meme pool*.

Memes manifest from thoughts or discoveries, and either survive in the meme pool or die off, depending on their value as perceived by living, behavioural individuals. Memes may also be represented in cultural patterns. A selection, perhaps, for a preference for a particular way of dressing, dining, or dancing, sends out cultural information arrays, which offer differing benefits to each individual that comes into contact with it. As this information feeds back between individuals and their groups, the effects spread, and new subcultures form.

Memetics, the study of memes, describes how fashions come and go in what are often compared to viral or contagious effects. Other examples of memetic patterns include words, songs, phrases, beliefs, trends, habits, and so on. Meme-gene coevolution, or *Duel-Inheritance Theory*, explains the ecology responsible for some of our ways of doing things and their genetic influences. Lactose tolerance, the Western acquisition of an adaptation involving increased tolerance to cow's milk, is just one example.

Other theories suggest stranger forces at play. Rupert Sheldrake, in studies from 1999–2005, showed that many people can sense when they are being stared at. His theory of morphic resonance proposes that similarly shaped ideas sometimes travel instantaneously between similarly shaped homes. Memes could be travelling between minds as quantum synchronicities in a entangled holographic universe.

Memes are everywhere, affecting many of your actions.

Like genes, they can survive intact for centuries.

Some memes express themselves in trendy dress or speech.

Others are injected into society by advertisers and politicians.

Memes can take over what we notice, and what happens.

Folk wisdom is a fine kind of meme.

Extraterrestrial Life
in an evolving universe

The universe is a very life-friendly place (*see page 376*) and there are over 100 trillion billion stars in just the part of it we can see, so there must be other planets out there with biological life. Surely the universe is teeming with evolved intelligent beings—so where are all the aliens? This odd conundrum is known as the *Fermi Paradox*.

There are many answers and here are a few of them. Maybe there are no aliens. Perhaps we are alone in the universe. Are aliens out there but we haven't met up yet? Maybe everyone is too far away, too unevolved, or too highly evolved. What if long-distance space travel is harder than we thought, even impossible? Could we be using the wrong technology, the wrong frequencies, the wrong eyes?

Look at what we've achieved as a species in the last 10,000 years, even the last 100—what does a species like us evolve into after another 1,000? Would we be even biological anymore? Physical? Perhaps we evolve into teleporting psychic quantum beings of pure light and conscious code information! Or maybe selfish careless lifeforms like us adopting selfish corporate economic models always self destruct. Maybe aliens aren't bothering with us because we're still just pond slime.

Then again, maybe aliens *are* here already. Life appeared so early on Earth (after just 600 million years) that perhaps it didn't evolve here at all, maybe it was seeded by bacteria trapped in cometary ice or space dust (a theory known as *panspermia*). Could we have neighbours, might we be children? Are we property of some kind? Are we being farmed? Watched over? Are we in quarantine? Dangerous keep out!

Perhaps the entire universe is itself evolving toward some future superconscious information-rich state we cannot yet imagine.

Above: An alien planet. Life might exists in sulfurous or silicate parallels to our version. Other types of nucleic acid structures could have conjured themselves into cells. Life could exist on moons like Jupiter's watery Europa or Saturn's Triton.

Above: Convergent evolution suggests that alien trees may look strangely like earthly ones, plant-grazers may resemble horses and squirrels, and their eyes will be very similar to ours. Shark-shaped things will swim in seas.

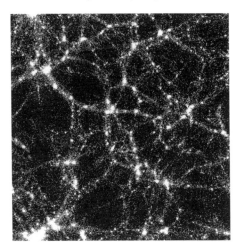

Above: Superclusters, each containing 50,000 or more galaxies strung like pearls along filaments of galaxy clusters that make up the macrostructure of the universe. At this scale supernovæ appear as occasional bright twinkles.

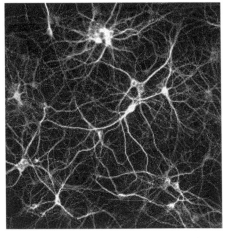

Above: Neurons in the human brain. Neurons fire electrical signals which travel down dendrite filaments. Could the entire universe be a vast quantum brain which somehow imagined its best self as it was being born?

BOOK V

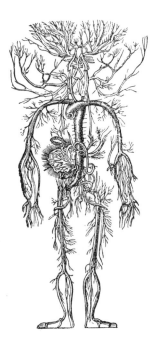

Two early engravings of human trees of veins and arteries. Tree-like structures are an
ideal way to maximize and connect points of contact within a finite space. We share
some genes with trees, and early geneticists thought that the number of chromosomes in
an organism determined its prowess, until they discovered that humans have the same
number as privet hedges, while mice have slightly more, and lilies 33 times as many.

THE HUMAN
BODY

Moff Betts

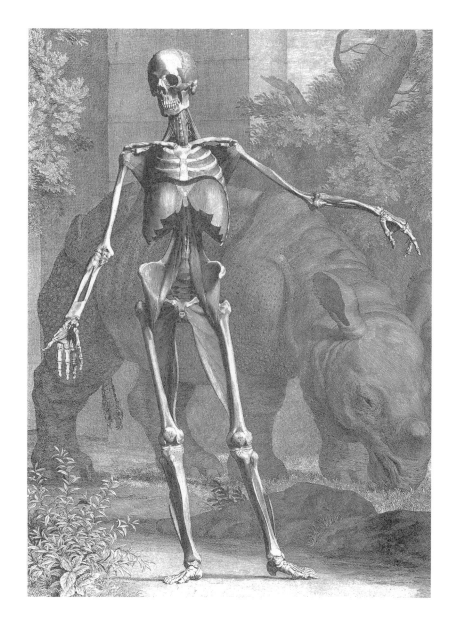

INTRODUCTION

The human mind is so curious, and our bodies are so intricate, that inquisitive men and women have made our own flesh the most intensively studied field of all. So when asked to write a tiny book about the human body, I protested that the topic was too large and complex to fit into such a small space. As it turned out, the book had ideas of its own, but is necessarily dense and in places a fairly simple version of what we know. With the combination of cram and scatter used in these pages, you may find that it all fits together better on a second read through.

What follows starts with some history of our investigations into the body, then proceeds to examine the microscopic mechanisms that have become the cornerstone of modern biology. The bulk of the book comprises a look at the functions of the major bodily systems, before ending with some examples of integration. If you are unfamiliar with the life sciences, the small print glossary-index at the back of this book may also be helpful.

Being human isn't all about constant awareness of our bodies, indeed some of our best experiences happen when we manage to forget about them altogether. Unique, doomed to death, and your only true possession, your body's frantic devotion to keeping you alive is roundly ignored most of the time, allowing you to get on with the brilliant business of life. But I hope that these few glimpses of our discoveries about the human body, many of them truly amazing, will leave you with the feeling of gratitude and awe that this enigmatic and unparalleled form deserves.

ROUND WINDOWS
see for yourself

Little was known of the interior layout, let alone the workings of the body, until remarkably recently. As Europe floundered in the Dark Ages, when cures like eating pigshit for pleurisy were in vogue, cosmopolitan Baghdad and cerebral Esfahan shone as centers of medicine, while India and China had, and still have, sophisticated systems from antiquity.

Tellingly, dissection was fairly taboo in all cultures until Renaissance Italy, where the anatomists who made the first systematic cuts named their discoveries, like Fallopian and Eustachian tubes, after themselves. Animal-based errors and blind conjecture, fabricated by Galen in second century Alexandria, ruled Western ideas of anatomy until Vesalius published the first definitive map of the body in 1543, informed by an ample supply of executed convicts.

Mankind's ensuing autopsy ('see yourself') has exploded us into a bio-machine of innumerable bits, mostly studied in dead humans or half-dead furry mammals, a far cry from a living unity.

The living human tissue you can see most clearly without cutting someone open is an iris; looking at someone else's avoids the left-right error of the mirror. No two are the same, even in one head. Iridologists can see your whole body mapped onto the iris, which is mainly muscle coloured by pretty opaque pigments shielding the light-sensitive retina at the back of the eye from over-exposure. Outer radial muscles dilate the pupil in the dark, balanced by an inner circle which constricts in the light, increasing depth of field. Translucent aqueous humour, filtered from blood by tendrils behind the iris, flows freely in its catacombs, even during rapid motion.

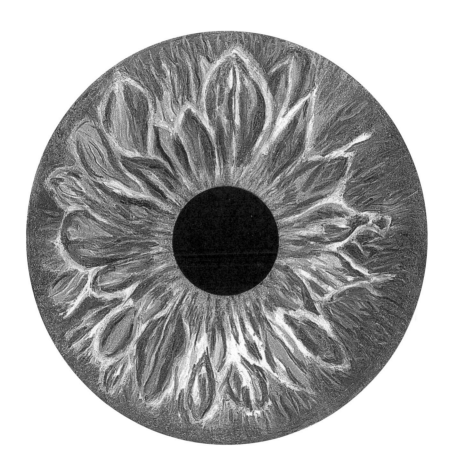

ASIAN FUSION
ancient Indian and Chinese systems

The basic premise of all ancient systems is that we are part of a larger living body, the cosmos, and are shaped by its forces.

In Chinese cosmology, unity divides into the opposites *yin* and *yang*, which resolve as a vortex of *chi*, energy that flows through all things. Chi's flux through humans was initially charted by observers who presumably had an idea of the likely internal apparatus from animal butchery and imagination. They sensed chi-channels on the body, named meridians, which map onto twelve internal organs, with central connexions out of sight. Meridians guide the flow of organ-specific chi around your innards and out through your extremities. Being alive is the ability to accept, transform, and radiate chi, in all its infinite guises, back to the universe.

Each of the 361 vortices along the meridians, often used as access points, has its own quality of chi, reflecting an aspect of its related organ, itself ruled by the interplay of five elements (*see page 397*).

India's two middle meridians form a double spiral which snakes up the spine, channelling *kundalini*. This is rising *prana* (chi) whose yin and yang are *shakti* and *shiva*. Where the snakes cross, seven midline vortices form *chakras*, spinning coloured two-way prana funnels which fluctuate according to your current state.

These perennial systems not only map inner states onto your skin, but also encompass the outer influences of human interfaces with nature, such as diet, habits, and locale, so your total level of integration with the cosmos determines your bodily vitality.

Baffled boffins interpret prana, or chi, as electricity, central nervous *gating* effects explaining the ability of acupuncture to anæsthetize.

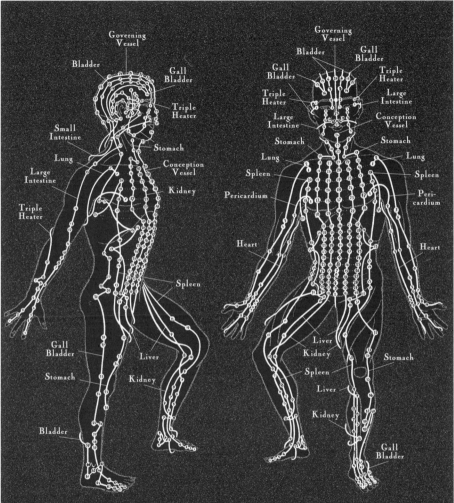

Each meridian is a pathway of acupuncture points forming a circuit with an organ via inner pathways. The twelve bilateral meridians then also form a continual circuit in the order Lung, Large Intestine, Stomach, Spleen, Heart, Small Intestine, Bladder, Kidney, Pericardium, Triple Heater, Gall Bladder and Liver, connected through additional internal meridians. The two central meridians bring the total to fourteen.

ELEMENTS AND HUMOURS
man mirrors nature

Medieval folk assumed they were composed of the same number of elements as nature, which was usually three, four, or five (*see pages 396-7*). Inner versions of fire, air, earth, and water swirled around as humours, blending in the body, and painting its faces, seasons, and qualities. Like the weather, humours had mild, stormy, heavy, and beautiful days inside each person. As climatic patterns sculpt local landscape, humours revealed the deeper colours of your 'complexion'.

Humours' carrier fluids were only visible in the raw during an illness —the body's brave attempt to restore humoural harmony. Too much humour (in the wrong place) could be cured by efflux of blood, phlegm, pus, or bile, thus freeing the intangible humour.

The Swiss mystic and iconoclast Paracelsus realized the ethereal humours must have physical counterparts, basing his system on an alchemical trinity of salt, sulfur, and mercury. As in nature, these combined inside the body, making fixed, mutable, and cardinal compounds, which conferred structure, energy, and communication, like the molecules now found in microbiology.

Subtle blending effects can also be seen in modern *endocrinology*, whose *hormones'* characters emerge starkly in various syndromes of excess or depletion, giving clues to their patterns in health.

Just four chemical elements (H, C, N, and O) combine in the four *bases* of *DNA*, the *molecule* that carries hereditary complexion. A fifth element, phosphorus, joins the pentagonal sugars of DNA's backbone, to make a long chain which has a mirror-image, the two strands resolving as the serpentine double helix (*see pages 193 and 230*).

hot

CHOLERIC
nervous - plasma - fire
hydrogen

SANGUINE
arterial - blood - air
nitrogen

dry

the

HUMOURS,

tempers,

elements,

&

other esoterica
phosphorus

wet

cold

MELANCHOLIC
digestive - bile - earth
carbon

PHLEGMATIC
lymphatic - mucus - water
oxygen

PRE-CONCEPTION
two generation gaps

The egg bobbed at anchor, with a thousand others, made in your mother's ovary before her own birth. A generation later it set sail, and met your father's sperm swimming into port, spawned just two moons earlier, as one of that day's batch of 300 million.

Apart from the fusion of *gametes* (sperm and egg) at the moment of conception, all cells are made by cell division; *mitosis (see page 196)*, or in the rare case of gametes, *meiosis*. A gamete has 23 *chromosomes*, the long strands of DNA that carry *genes*, so when two gametes unite, the conceptus has 46, which are copied into every cell in your body. The chromosomes in the gametes that came together to make you were remixes of your grandparental chromosomes, made in prophase I of meiosis, when each of your parents' parents' pairs swapped parts of their limbs. So you are a meeting of your parents, but a proper mix of your grandparents.

Girls' 23rd chromosomes are Xs from both parents, but boys inherit their father's Y and one of their mother's Xs. Of their two Xs, girls inactivate one in every cell to form a *Barr body*.

Vanguard sperm batter the giant *ovum*'s shield before one lucky straggler gets in, bringing its 23 chromosomes to join the 23 that the ovum doesn't discard at this point (she waited for ages in late anaphase II). This union creates a unique new set of 46 chromosomes.

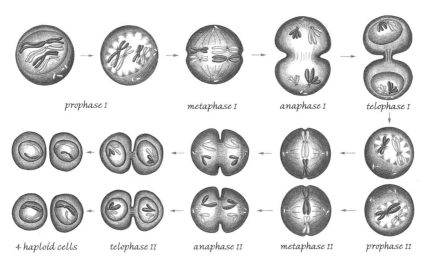

prophase I   metaphase I   anaphase I   telophase I

4 haploid cells   telophase II   anaphase II   metaphase II   prophase II

MEIOSIS: Unique new eggs and sperm are created from parental DNA.

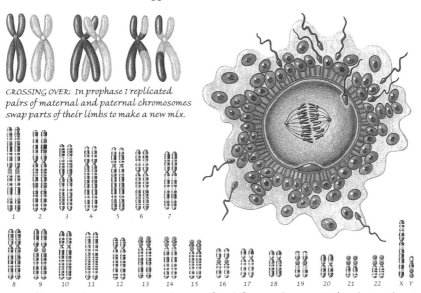

CROSSING OVER: In prophase I replicated
pairs of maternal and paternal chromosomes
swap parts of their limbs to make a new mix.

1   2   3   4   5   6   7

8   9   10   11   12   13   14   15   16   17   18   19   20   21   22   X   Y

A set of human chromosomes, 23 pairs, one of each from each parent (a boy, last pair XY).

AND DNA
cooking up shapes in the protein kitchen

Down in molecule sea, DNA functions as a cookbook. Written in a four-base 'alphabet', *a c g* t, our six billion base *genome* is 97% rhythmic, mainly long *tata*-rich passages. Amidst this babble the odd coherent recipe for a 'shape' pops up, as a readable string of three-letter *codons*, like *tac-gtc-gag-cct-cac-gtt-cat-ctc-taa-ttg-ccc-atc*.

The recipes are for chains of *amino-acids*, which form *proteins*, the next level of molecular shape and function. DNA also holds recipes for *RNA*, her busy single-stranded, single-use sister, who runs the kitchen and makes all the utensils. She copies recipes, chops and peels, and builds multi-protein dishes, while dreamy DNA just lets herself be read, replicated, and rarely modified.

DNA's two helixes are mirror images, because the bases pair off, *a-t* and *c-g*, linked by *hydrogen bonds* across the spiral innerspace. RNA reads one thread, the back-up side coding for any necessary repairs, and enabling the twin zipper to *supercoil* into a superhelix. A gene is a set of recipes coding for a finished protein (or RNA), and 'switches on' when all segments uncoil, unzip, and are read.

Most of the time DNA lies partially coiled inside the *nucleus*, only assuming the familiar 'X' shape during cell division, when two copies of a replicated chromosome are still joined at the hip.

Knowing how short one life is, DNA hops across chromosomes, species, and generations. *Viruses* hijack your kitchen, while *bacteria* pass on their crazy new recipes, helping evolution along. Many of our recipe ideas derive from evolutionary pioneers like seasquirts, lungfish, and millipedes, to whom we owe much of our structure.

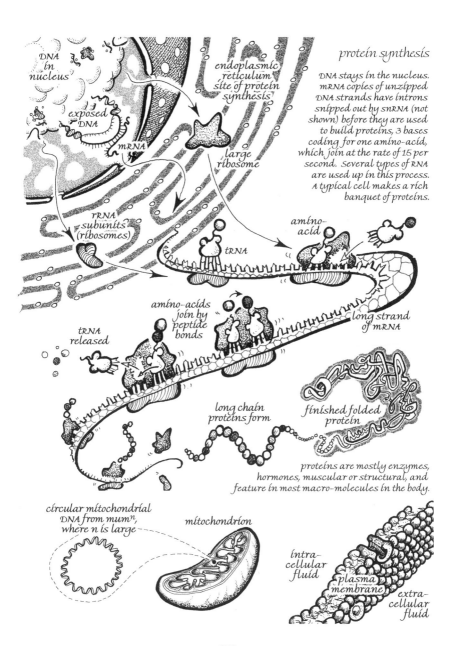

protein synthesis

DNA
in
nucleus

exposed
DNA

mRNA

endoplasmic
reticulum
site of protein
synthesis

large
ribosome

DNA stays in the nucleus.
mRNA copies of unzipped
DNA strands have introns
snipped out by snRNA (not
shown) before they are used
to build proteins, 3 bases
coding for one amino-acid,
which join at the rate of 15 per
second. Several types of RNA
are used up in this process.
A typical cell makes a rich
banquet of proteins.

rRNA
subunits
(ribosomes)

amino-
acid

tRNA

amino-acids
join by
peptide
bonds

long strand
of mRNA

tRNA
released

long chain
proteins form

finished folded
protein

proteins are mostly enzymes,
hormones, muscular or structural, and
feature in most macro-molecules in the body.

circular mitochondrial
DNA from mum^n,
where n is large

mitochondrion

intra-
cellular
fluid

plasma
membrane

extra-
cellular
fluid

THE EARLIEST SURVIVOR
one stormy night about four billion years ago

Organic life sprang from one first cell, formed amid a swarm of molecular mosaics in a long lost pool. Just how all the ingredients coalesced in the right circumstances, nobody knows, but happily, the mother of us all survived, replicated, diversified, and evolved.

After the dark early eons, cells used light, carbon dioxide (CO_2), and water to make carbohydrates and oxygen, as plants still do. With atmospheric oxygen rising wildly, one cell mastered a chain reaction that used this fiery gas to turn carbohydrates and water into CO_2 and energy, as we still do. This genius made it welcome inside other cells, thus furthering the tradition of *endosymbiosis*. Hundreds of descendants of these *archæbacteria* live inside each of your cells, as *mitochondria*, and the air you breathe is for them.

Pro-mitochondria weren't the only cohabitants in early cells, but have kept their genes and reproduction largely to themselves. *Eukaryotic* cells, which make up complex organisms like plants and humans, were made by several organisms, with pooled genes evolving into a nucleus. So our 'smallest living subunit', the cell, itself began as a cooperative of even smaller earlier organisms.

Prokaryotes (bacteria) retain the ability to exchange genes (*below*) just as we swap ideas, the global bacterial superorganism accessing a common gene pool in order to adapt to local niches.

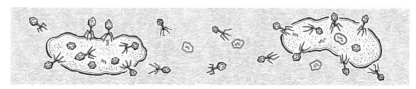

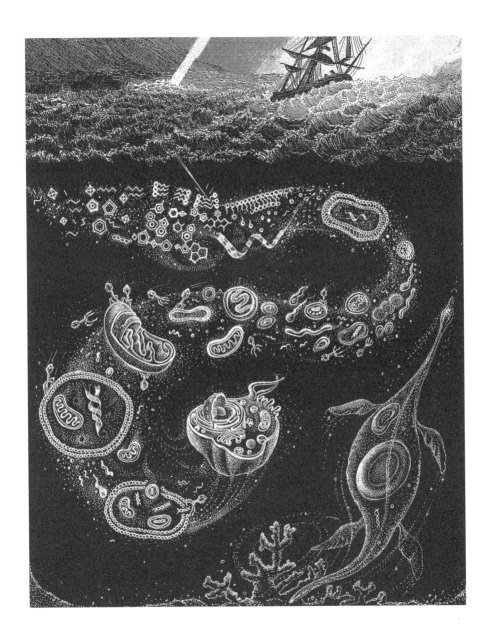

THE CELL
life's electric microcosm

You are made of cells, many trillions of them, in forms as diverse as long delicate nerves to cuboid mucus factories, each of which performs a certain function as its contribution to the whole.

Each cell differentiates into one of about 300 specialized types by switching on only a few of your genes, so makes only certain proteins, which engineer other proteins, fats, and sugars to build the cell's structures, which define its function. The three-dimensional diversity of proteins is what enables their complex molecular tinkering and the shape-specific interactions crucial to messaging systems.

Cells' internal soup differs radically from the extracellular fluid, operations taking place inside the thin fluid *plasma membrane*. This is studded with protein-based structures, mainly *channels*, governing what enters and leaves, *receptors*, transducing signals from other cells, or *pumps*, like the ubiquitous *Na-K-ATP pumps*, which swap sodium (out) for potassium (in). The flux this sets up makes cells electrified, negative inside and positive outside.

These pumps use about half of all your energy, which is basically your available *ATP*, made by mitochondria with a proton-electron transfer chain that uses food and oxygen to fuel ATP synthesis.

Cell division, *mitosis (below)*, happens at intervals from every few hours in high-turnover tissues like front-line white cells (*see page 298*), to probably never again, in the case of most of your brain cells.

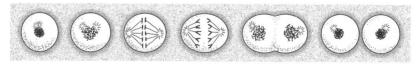

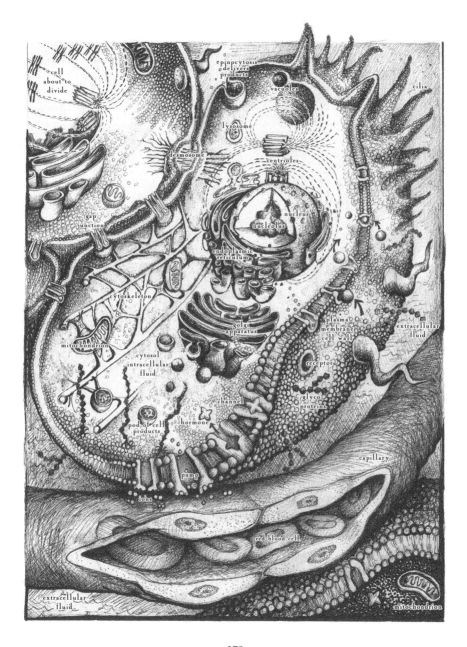

cell
about to
divide

pinocytosis
delivers
products

vacuole

cilia

lysosome

desmosome

centrioles

nucleus

nucleolus

gap
junction

endoplasmic
reticulum

cytoskeleton

golgi
apparatus

plasma
membrane
cell wall

extracellular
fluid

mitochondrion

cytosol
intracellular
fluid

receptor

channel

glyco-
proteins

pod of cell
products

hormone

capillary

pump

ions

red blood cell

extracellular
fluid

mitochondrion

TISSUES
saltwater archipelagos

Tissues are communes of similar cells performing the same tasks; several make up any organ. Tissues fall into four major classes:

Epithelial tissues are highly selective about what they let in and out, forming linings of hollow and tubular organs; many make a mucus matrix to protect and extend the border they control.

Connective tissue includes wandering and static cells; they deposit artefacts outside their walls to build ultra-structures and general meshwork, as well as defining your immune boundaries.

Muscle comes in three types: Skeletal muscle contracts linearly under conscious control; smooth muscle squeezes circularly, narrowing the tubes in whose walls it writhes, subconsciously guided by *autonomic* nerves; and heart muscle's tireless cells also pass on electricity to their neighbours, at differing speeds in order to synchronize the pattern of cardiac contraction. Apart from a few nerve cells in the guts (which have a mind of their own) all nerves are linked in one giant electrical system.

Like Gaia, and life in general, we are two thirds live saltwater, which bathes and joins all tissues. The three primary fluids are: intracellular fluid, which varies according to cell; extracellular fluid, the sea around cells; and blood (*below*), half cellular (mostly red *corpuscles* carrying gases) and half fluid, *plasma*, carrying various minerals and molecules between the different tissues.

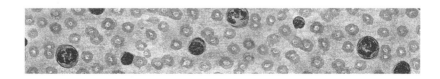

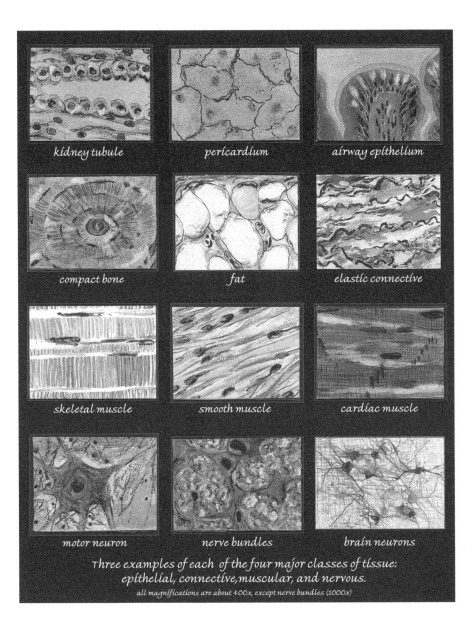

kidney tubule

pericardium

airway epithelium

compact bone

fat

elastic connective

skeletal muscle

smooth muscle

cardiac muscle

motor neuron

nerve bundles

brain neurons

Three examples of each of the four major classes of tissue:
epithelial, connective, muscular, and nervous.

all magnifications are about 400x, except nerve bundles (1000x)

EMBRYOLOGY

one cell becomes human

Uterine milk nourishes the embryo during the rapid cell division of the first few days. On day five the bundle of 32 cells bores a hole in its coat and implants into the wall of the womb, which feeds it, helped by the yolk sac, until the placenta, made jointly by mother and baby, is fully functional by the end of the first month.

Three primordial 'germ' layers (*below*) develop into all tissues: ectoderm forms brain, nerves, and outer skin; mesoderm makes solid organs, flesh, and bone; endoderm forms most epithelial tissues.

The influences that make embryonic cells specialize and grow in patterns are still secret, but guidance seems to come through prenatal master genes, local signal gradients near the developing blood vessel tree, and streams of pilot cells that die before birth.

The busy infolding, migration, and reunion of diverse pockets, arches, and tubes of cells proceeds so quickly that by two months it has made nearly all its little organs, and is recognizably human.

It just remains to make some gonads. In boys, a signal cascade in week seven is started by the Sry-1 gene on the tiny Y-chromosome, which acts on some cells that crawled from the yolk sac to the lower pole of the developing kidney, and results in male genitals, with all their hormonal consequences. In girls, X-driven maturation continues, undiverted, to form womankind.

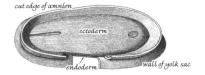

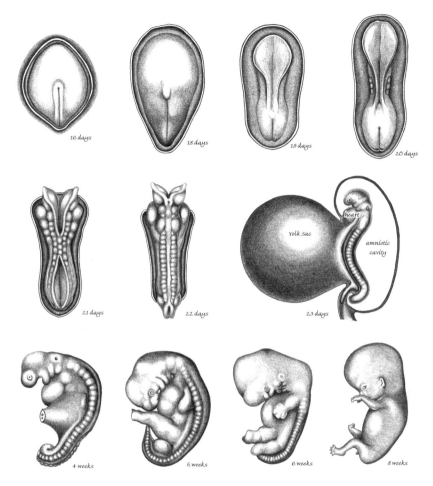

16 days   18 days   19 days   20 days

21 days   22 days   23 days

heart

yolk sac

amniotic cavity

4 weeks   5 weeks   6 weeks   8 weeks

Early cells are totipotent, meaning they can become any type of cell, but gradually lose this ability as they differentiate into germ layers, then tissues and organs. A midline stripe, the early nervous system, first gives the embryo a head and tail, and the fastest- developing early organs are the brain and heart. Five arches migrate from the top of the heart to form head and neck structures, like ears, cranial nerves, and the thyroid, the earliest endocrine gland. Buds from the primitive gut form the lungs and liver, which begins making the first blood cells. The lower body develops from 44 paired segments which give rise to bones, muscles, nerves, and inner organs. Limb buds are forming by seven weeks, as is the young face. After eight weeks, growth rather than differentiation takes over.

THE DISSECTOR'S DREAM
systems and organs

Three truths about your body are it was once one cell, it's alive, and it's an organized, integrated whole. After that, any analysis divides it into subunits, be they cells, tissues, organs, regions, or chakras, all defined by specialization of form and function.

Western physiology identifies sets of cooperative systems, each performing a major task, like blood circulation. Any given system is made up of contributions from several organs, which themselves can be localized collections of tissues, like the heart, kidneys, or intestines, or diffuse like blood vessel or nerve trees.

Any tissue contains a specific array of cells, whose vast range of specialization makes it possible for huge beings like us to exist. Bathing in the warmed and modified sea we have trapped in our skin, cells are grateful slaves, told what type to become in embryonic life, then controlled by signals from elsewhere.

All the body's systems are completely interdependent, and all intercommunicate, to varying extents, with the brain sensing and controlling the activity of the whole body, receiving and sending signals as short-term nerve impulses, or longer-term hormones. The brain can override any other processes trying to happen, especially in adults, for example Tibetan monks who master the art of sweating while naked in a snowstorm.

The cartoon systems depicted opposite mostly have their own pages later in this book, but like a single page of a story, each would be meaningless considered in isolation, which dissectors sometimes forget. Much is still unknown, for instance how we can imagine the taste of our ideal food or herb of the moment.

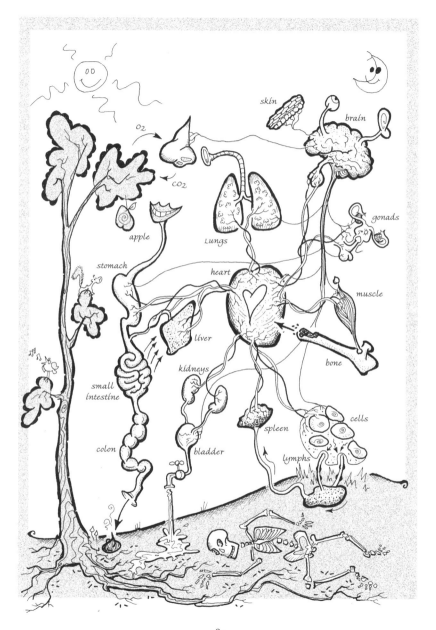

FOUR PLUS ONE
is phive

Variations in human shape are started off in utero, and the three very basic shapes can be traced to the relative dominance of one of the three embryonic germ-cell layers, and their tissue destinies: ecto- (wiry), endo- (rounded), and meso-morphs (muscular) (*see page 396*).

The first thing a baby recognizes visually is the human face, and then its stunning variability. Within the obvious sameness of the human form are some less evident constants. Phi, ϕ, the golden section, which rules organic growth and pentagrams, defines the change of your center during growth. A baby's midpoint is his or her belly button (the past), with the genitals a golden section away from the crown, but by adult life, the genitals (the future) are central, with the belly button ϕ from the feet. The relative lengths of the bones in your arm and hand display a golden series, and ϕ also crops up in the geometry of DNA (*opposite*).

Humans, like DNA, show four-plus-one patterns, as four limbs and a head protrude from the torso, and hands sprout four fingers and a thumb. Like fingers and toes, children have five teeth in each quadrant of their mouths; eight more per quadrant develop and emerge later. The total amount of teeth you chew your way through in one life is thus 20+32, a full deck of 52.

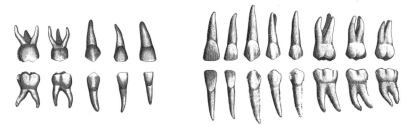

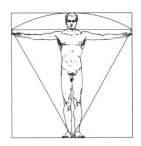 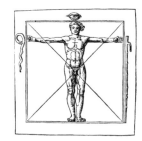 

Span is equal to height in an average adult, with the gonads halfway up or down. In a young child the belly button is in the middle. The Golden Section, ϕ, marks the position of the belly button in adults and the gonads in children.

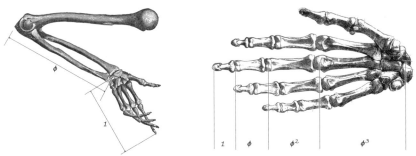

The primary bending places from fingertip to elbow relate to their neighbors by the Golden Section, here 1.618, found in pentagrams and the water molecule.

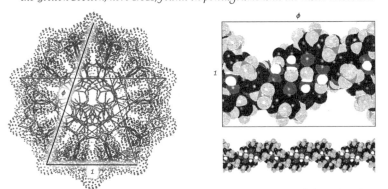

A DNA cross-section shows the 10-steps-per-spiral nature of the molecule; A single twist of DNA fits a Golden Section rectangle.

EARTH AND FIRE
transformation in gut and liver

The pleasant thought of food, which makes you go and find some, also stimulates the secretion of digestive enzymes from the pancreas, and bile from the gall bladder. These break down food into microscopic morsels in the 'small' intestine whose *villi (below left)* absorb them over a surface area the size of a tennis court.

Nutrients are intelligently transported into villus cells, and enter the blood through *capillaries*, which feed this rich soup, via the portal vein, to the liver. Here it mixes with fresh arterial blood in the porous cells of liver lobules *(below right)*, where molecular wizardry transforms whatever arrives into whatever you need, for instance making sugars out of proteins, and renders all sorts of poisons harmless or usable. Most nutrient distribution proceeds from the well-named liver. Prometheus, chained to the rock of earthly life, had his pecked out every day by a sea-eagle. By night it grew back to full size. Likewise, the liver sacrifices its resources and integrity to meet the bodily demands of each day, and by night it restores order, mends itself, synthesizes new supplies, and makes ready for the sea-eagle that comes with the sunrise.

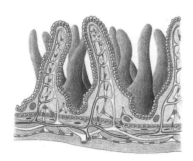 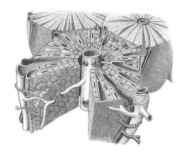

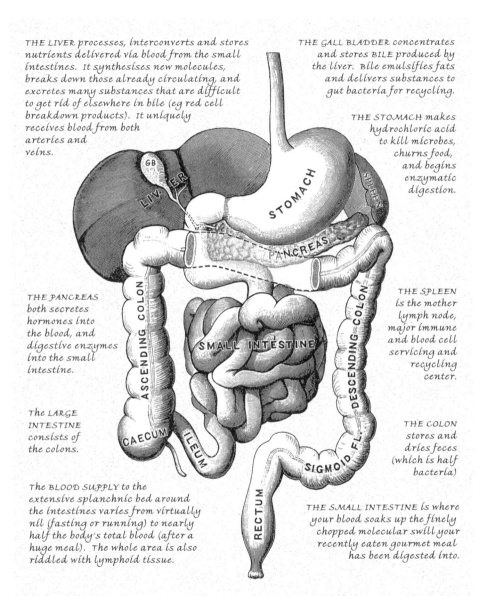

THE LIVER processes, interconverts and stores nutrients delivered via blood from the small intestines. It synthesises new molecules, breaks down those already circulating, and excretes many substances that are difficult to get rid of elsewhere in bile (eg red cell breakdown products). It uniquely receives blood from both arteries and veins.

THE GALL BLADDER concentrates and stores BILE produced by the liver. Bile emulsifies fats and delivers substances to gut bacteria for recycling.

THE STOMACH makes hydrochloric acid to kill microbes, churns food, and begins enzymatic digestion.

THE PANCREAS both secretes hormones into the blood, and digestive enzymes into the small intestine.

THE SPLEEN is the mother lymph node, major immune and blood cell servicing and recycling center.

The LARGE INTESTINE consists of the colons.

THE COLON stores and dries feces (which is half bacteria)

The BLOOD SUPPLY to the extensive splanchnic bed around the intestines varies from virtually nil (fasting or running) to nearly half the body's total blood (after a huge meal). The whole area is also riddled with lymphoid tissue.

THE SMALL INTESTINE is where your blood soaks up the finely chopped molecular swill your recently eaten gourmet meal has been digested into.

AIR AND WATER
balance in lungs and kidneys

In and out endlessly like the tides, the air you breathe refreshes the blood with oxygen to take to mitochondria, while removing the CO_2 they synthesize. Blood gas levels are monitored by the *brainstem*, which adjusts the rhythm and depth of inspiration.

Bronchi branch about 25 times, ending in bundles of elastic sacs named alveoli (*below left*), with a surface area, like villi, of about $80m^2$, over which air exchanges gases with *hæmoglobin* inside red blood cells, flowing at twice the usual speed, in the lung capillaries.

Kidneys' *glomeruli* (*below right*) have basal structures also found in alveoli. While the lungs expand to deal with gas balance, the kidneys head inward to regulate fluids. Kidneys take as much blood as the brain, and send signals which regulate blood pressure and make you thirsty. They filter the blood, then reabsorb 99% of the filtrate, to fine tune your overall fluid composition.

Of the lungs' blood supply, a Syrian *Qanun* (*ca.* 1250) says,

"When the blood has become thin, it is passed into the artery to the lung, to mix with the air. The fine parts of the lung then sieve into its vein, which arrives at the left of the two cavities of the heart."

 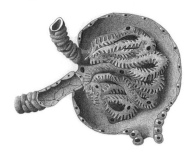

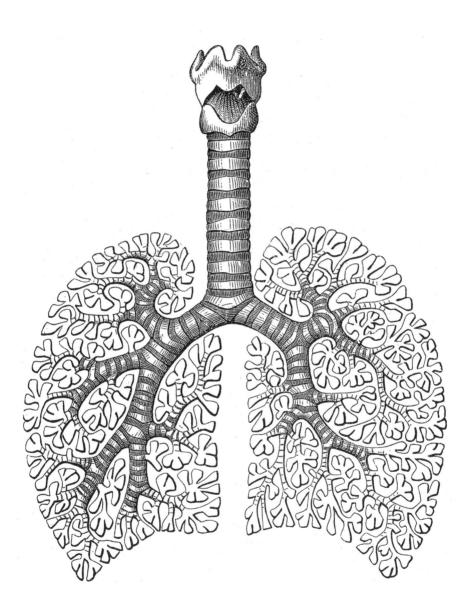

THE HEART
two beat as one

The heart starts life as two primitive tubes that unite to form a cylinder, which is beating by the end of the embryo's third week. It grows, twists, and segments into a four chambered pump which beats non-stop about a hundred thousand times a day.

Each heartbeat is an electrically fired synchronized contraction which initiates in a node of *pacemaker cells* in the right *atrium*. The pulse is differentially conducted over the heart to become a rising spiral wave. The easily detected electromagnetic field of this is often shown in two dimensions as ⩜. In each beat, the two atria squeeze pooled blood through the *atrio-ventricular* valves into the *ventricles*, which wring it out in a vortex through the two *semilunar* valves. The lub-dup sounds are the four valves shutting, in two pairs.

This twin pump contracts as one, but runs a pair of circuits. The right heart collects blood from the body and sends it to the lungs, from where it returns, freshly aired, to the left heart, which pumps it around your body again, including to itself, via the frugally narrow coronary vessels. Half of all *bloodflow* at any time goes through your lungs, and at the middle of this endless air-body loop is the heart, which thus unites the interior with the exterior.

Heart cells transmit electricity to each other, like nerves, though unlike the brain, the heart has only one basic electrical pattern. The electromagnetic field of each beat pulsing through the brain's sensitive nerves may be why some say the heart is the deepest core of the mind. The *cockles* of your heart are warmth-sensitive organelles whose anatomical location has yet to be discovered.

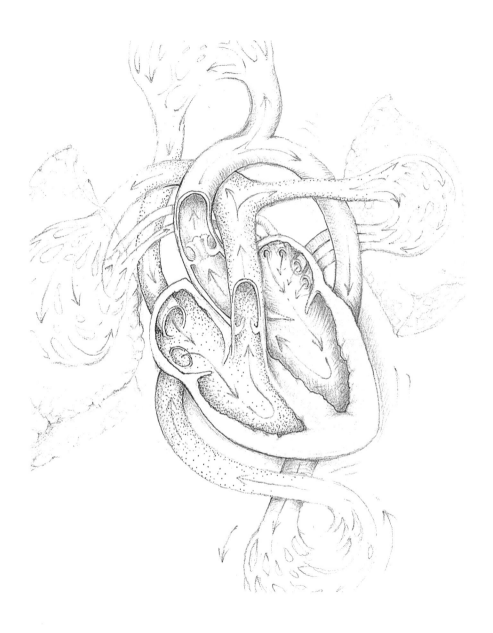

VEINS AND ARTERIES

two trees unite

The fact that the blood circulates endlessly was not discovered until 1628 by the no-nonsense Englishman William Harvey.

The arterial tree from the left heart branches to supply the body with blood, carrying oxygen, food, and signals. Between the blood and tissues is the micromesh of capillaries, tiny tubes that a red cell can just squeeze through, where gases, molecules, and ions soak out through the thin capillary walls. Here the blood picks up cell produce, and waste, including CO_2, before a twin tree converges into veins, returning old blue blood to the right heart.

Blood cells are bred in the bone *marrow*. Red blood 'cells' have no nucleus, being disposable discs that maximize surface area for gas exchange. The molecule that carries gases inside red cells, and gives blood its colour, is hæmoglobin, whose four protein-pigment subunits each contain one iron atom. Each subunit carries a fiery oxygen in its iron-clad cage, and cleverly swaps O_2 for CO_2 according to local conditions in the tissues and lungs.

Red cells are loaded with *ATP*-generating systems that don't use oxygen, so don't use up their cargo. Unlike nearly all other cells, their outer surface is electro-negative, attracting them to the cells they supply through the extracellular fluid. So, an electric current flows in your blood vessels, with the heart as the battery. Artery-vein pairs often run with a nerve, the three sharing a name.

One red cell lasts about four months before it is chomped by a *macrophage*, in the spleen or liver, at the same rate as your marrow makes new red cells, which is about three million per second.

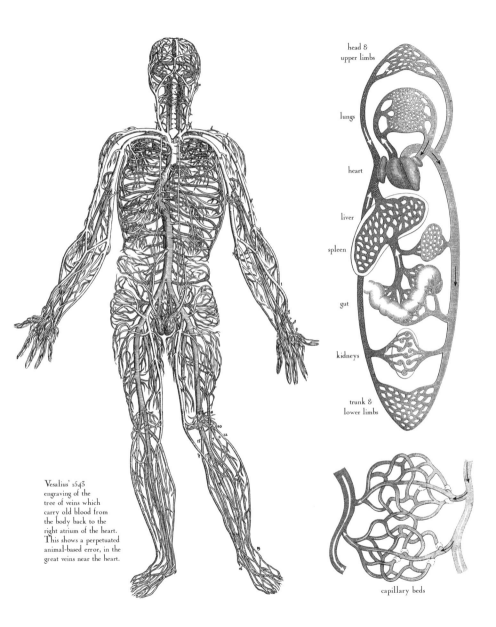

head &
upper limbs

lungs

heart

liver

spleen

gut

kidneys

trunk &
lower limbs

Vesalius' 1543
engraving of the
tree of veins which
carry old blood from
the body back to the
right atrium of the heart.
This shows a perpetuated
animal-based error, in the
great veins near the heart.

capillary beds

LYMPH PATROL
apprehension of miscreants

Like the sea, extracellular fluid contains flotsam and jetsam, so you have a lymphatic tree to soak up the overflow and detritus from around cells, and screen it for undesirables. Pure, filtered lymph is finally delivered back to the great veins near the heart.

White 'blood' cells are generated from the same stem cells as red cells, as are *platelets*, which plug any gaps in vessel walls. White cells run your immune system, and don't stay in the blood, but venture out, looking for trouble in the lymphoid network.

Where you touch the outer world is where grit and microbes get in, so gut, lung, and skin teem with colonies of lymphoid tissues. Here, various sorts of white cells come for brief or long stays, and messages flurry between them anywhere they gather, in order to coordinate activity, and alert each other to damaging intruders.

The most mobile of all cells is the *lymphocyte*, which squeezes through gaps it makes between capillary cells, gropes around in tissues, sampling forms (*see over*), then crawls into the blind-ended lymph ductules, to communicate its findings to other white cells.

Lymphocytes stop off and collaborate in nodes along this tree. Swollen lymph nodes ('glands') are only doing their job, packed full of lymphocytes, macrophages, and other types of white cell. Macrophages evolved from *amœbæ*, and engulf and digest things. Some innately recognize things like soot, others eat anything dubious-looking, and show chewed up bits of it to lymphocytes, whose verdict on their potential hostility will inform other white cells to destroy and recycle whatever needs eliminating.

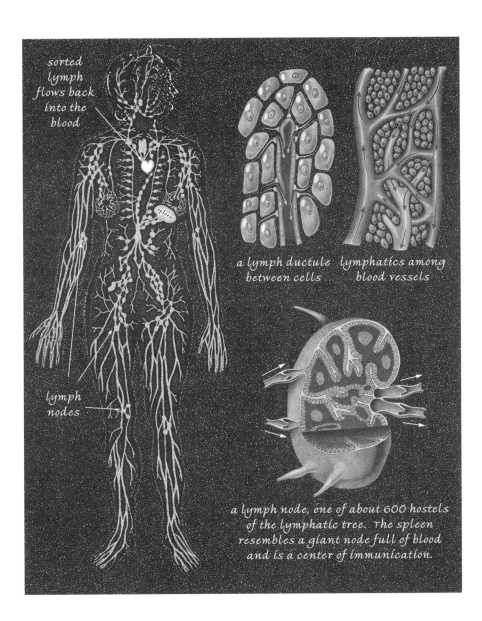

sorted
lymph
flows back
into the
blood

spleen

lymph
nodes

a lymph ductule
between cells

lymphatics among
blood vessels

a lymph node, one of about 600 hostels
of the lymphatic tree. The spleen
resembles a giant node full of blood
and is a center of immunication.

IMMUNOLOGY
discrimination of self

Childhood is when you learn the most, and nowhere more so than in the *thymus*, the gland that sits close to the heart. It spends your prenatal months killing any lymphocyte that gets alarmed when it encounters your own cell walls, which is most of them. So, by the time you are born, your few surviving lymphocytes only react to unfamiliar forms that are deemed to be not of you.

The shapes that identify cells as yours are mainly *MHC* proteins, and lymphocytes brandish other proteins which dock with these. If docking falters, the lymphocyte smells a rat, and sends messages that amplify far and wide to bring the white cavalry. To avoid whimsical wars being waged, several white cells agree to arouse each others' martial sides before letting full hostilities commence.

Your next weapons are either custom-cloned granulocytes which recognize and kill certain microbes, or antibodies, bespoke proteins produced by lymphocytes in response to an antigen. Antibodies co-operate to form attack complexes, or bind to microbes while signalling 'eat me' to a macrophage. Few intruders are hostile, and many ride along helping out, or pretend to be you in sneaky ways.

Veteran white warriors remember their battles, and can fashion weapons at short notice when the progeny of old enemies return. They are trained during childhood infections, when the immune army is learning military tactics. The thymus shrinks after your first birthday, and by dotage it has all but been replaced by fat cells. So as the years roll by, the school of discrimination between self and non-self gradually fades, a thymic idea of what one life is.

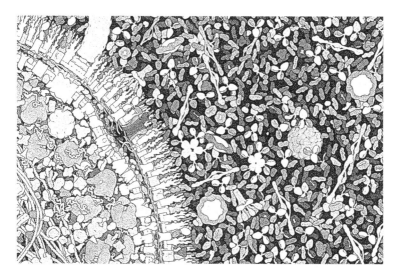

Above: an IgG antibody-C1 complex binding to a flagellated E. coli bacterium (left), triggering a cascade that leads to an attack complex piercing the intruder's wall. Below: a happy lymphocyte samples you, while several cells cooperate to wage war.

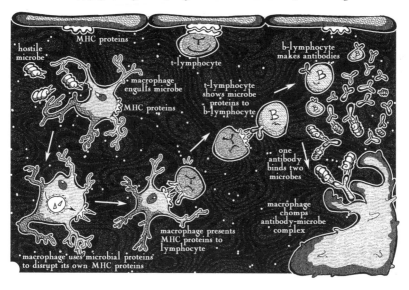

NERVES
your electrical intranet

The major divisions of the nervous system are *sensory*, *motor*, and *central* (the brain and spinal cord). Sensory *dendrites* bring data from the five peripheral senses into the central nervous system, where all the processing, integration, and thought take place.

Apart from a few *ganglia*, all nerve cell bodies, *neurosomes*, live in the brain and spinal cord, each supported by several *glial* cells, the space between the stars. Motor *axons* extend to fire muscles, and are 'voluntary', to skeletal muscles, or *autonomic*, subconsciously controlling smooth muscles in your tubes, glands, and organs.

The *voltage gradient* across nerve cell membranes is similar to that at the center of a thunderstorm (*see page 279*), and it fluctuates both rhythmically and in response to incoming nerve impulses. When the sum of local voltages reaches a threshold at the *axon hillock*, the cell fires, and a wave of temporary voltage gradient reversal thrills along the axon, as successive *voltage-gated* channels open to let sodium ions flood in for a millisecond. The impulse moves at walking pace in slow sensory nerves, and up to 250 mph as it hops along sleeves of *myelin* around the fastest motor axons.

Axons terminate at *synapses*, where a *neurotransmitter* is released, which travels to the next cell, usually to another nerve's dendrite, or to a muscle, or rarely to another axon. Here it makes the muscle contract (*see page 294*), or excites or *inhibits* the next nerve cell's frequency of firing. All nerve cells fire regularly, rate and rhythm defining their signal in the circuitry, and patterns of inhibition are what carve meaning in the white noise of the nervous system.

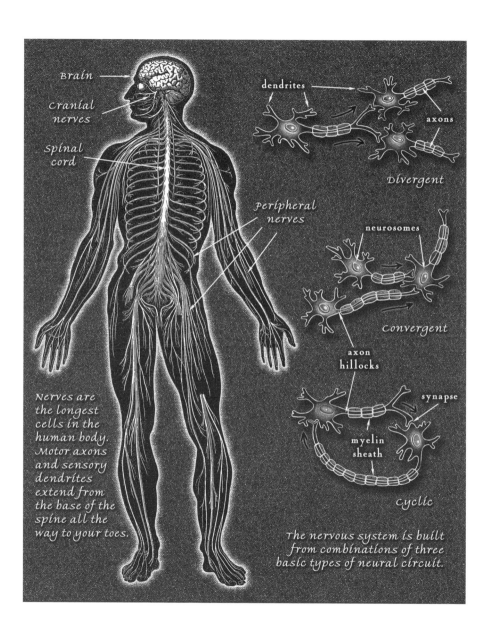

Brain

cranial nerves

spinal cord

Peripheral nerves

dendrites

axons

Divergent

neurosomes

convergent

axon hillocks

synapse

myelin sheath

cyclic

Nerves are the longest cells in the human body. Motor axons and sensory dendrites extend from the base of the spine all the way to your toes.

The nervous system is built from combinations of three basic types of neural circuit.

THE BRAIN
nobody's cracked the walnut

Intelligence had to find a structure on which to discriminate itself, and it went for the brain. *Homo*'s uniquely hefty walnut (only dolphins come close) localizes its subroutines so much that you even have a special area to hold images of your loved ones, while keeping their names and maps of where they are in other zones.

Given the countless electrical processes going on while you have one train of thought it is astonishing that we are capable of unified consciousness. Awareness seems to flicker on a 3 mm thin grey outer layer of *cortex*, which is all neurosomes, while the bulky *white matter* is all connexions. The musical right hemisphere and logical left converge on the midbrain, where conscious thought merges with subconscious control and integration processes, while your deepest feelings, instincts, and memories swirl around the *limbic system*. Most nerves swap sides through the brainstem, so the left brain corresponds to the right body, and vice-versa.

As its control center, the brain holds holographic maps of all the body's patterns, in diffuse and local forms (*see opposite*), in the same way as your mind holds inner maps of your outer world. The relationship between the brain and the body could be said to be similar to that between the human mind and the outer world, in that they both contain a clear idea of the whole in one.

Brainwaves vary with mental phases. Alpha and beta waves hum during daily life; spaced-out delta waves visit the young and the sleeping. Fairytale theta waves imprint childhood experiences, and only resurface in daydreaming and very emotional adults.

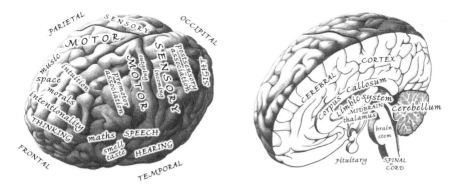

Above left: some local cortical functions. Above right: the corpus callosum is a bundle of connexions between the sides. The limbic system comprises peri-midbrain structures, like the hippocampus, and decides how significant or memorable you find things.

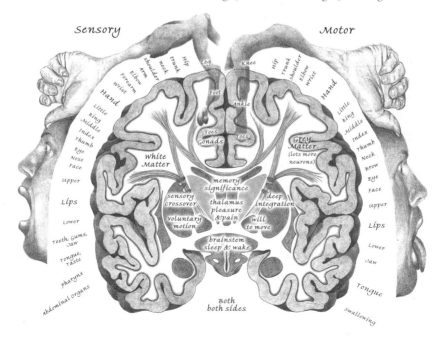

FIVE SENSES

and the twelve cranial nerves

All but XII *cranial nerves (see glossary)* run to and from the brain in the spinal cord. Smell, I, uniquely relays direct to the cortex, which says 'jasmine' or '?', as associated limbic circuits remember aromatic significance.

Retinal *rods and cones* turn off when light hits them, having been inhibiting bipolar cells, who flash the pattern of light and colour down II. Two inverted images meet at the optic chiasm, split and half cross, and flip twice more before radiating onto the occipital cortices, which integrate to show you a single vision. Most of your brain then decides what it means, and what to do about it.

III, IV, and VI swivel and focus the eyes. III also constricts the pupil, opposed by *sympathetic* nerves rising from ganglia in the chest.

V motors most of the chewing, and is the only cranial nerve to sense touch, via the dendrites in the skin of the face. Inner head spaces are under V, VII, and IX's intermingled sensorium, whose embryonic crossover explains tooth-ear confusion, and some of the weirdness of sneezing. VII also makes you drool, smile, and cry, and tastes the sweet, sour, and salty front of the tongue; the bitter back is IX's territory. The taste cortex warns the wanderer ('vagus'), *parasympathetic* X, of what may be heading the way of the innards.

Soundwaves beating the eardrum transmit via three bones to the oval window of the three-chasm *cochlear*, whose cells are tensioned to resonate in tune. The spiral harpsong travels down VIII, along with gravity and motion data sensed by the three-dimensional wing of the ear's bony labyrinth, which is thus crucial for dancing.

XII's fine motor control of the tongue and larynx lets you speak, and XI (*not shown*) keeps your head up and shrugs your shoulders.

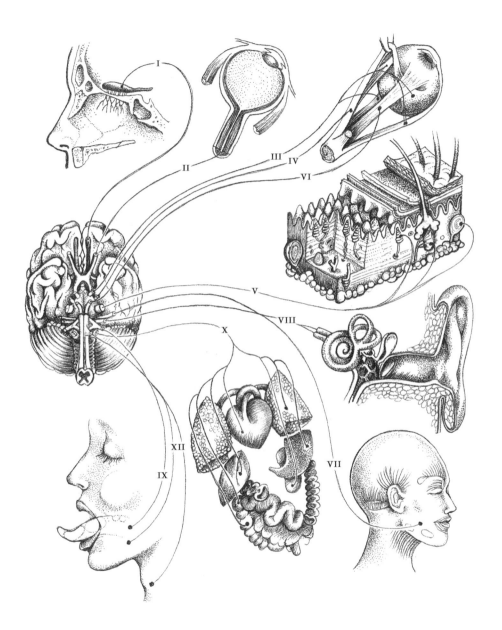

MUSCLE AND BONE
a spring-loaded calcium dump

What to do with heavy, double-charged, cell-trashing calcium? Dump most of it as shells, coral reefs, or endoskeletons in our case, and use the remaining few ions to tell it to hop and skip.

Calcium by nature alters the shape of molecules, so is ideal for signal transmission, while energy-dependent processes involve phosphate transfer. In muscle cells phosphates are pre-loaded onto an ATP between two overlapping proteins, before they contract. After the decision to move a muscle is made, every step in the cascade from brain to ATP uses calcium triggers, which finally unsprings the phosphate, and the muscle moves some tissue, blood, or bone, itself mainly crystalline calcium phosphate.

Your skeleton has 200 bones, plus three in each ear, where the tiniest muscle, stapedius, tensions the smallest bone, stapes.

Sensory endings in muscle called *proprioceptors* fire at a rate that depends on stretch, telling your brain the length of each muscle, so you can move accurately even with your eyes shut. The loop delay in this sensory-brain-motor circuit causes the 10 Hz tremor you can see when you try to hold your hand still.

Muscles contract in groups and sequences, controlled by the motor cortex and cerebellum where nerve cells fire in patterns to puppeteer complex routines, like writing or tumbling. These run more smoothly when not hampered by cortical overlay, also known as worry. Each cerebellar Purkinje cell has up to 100,000 dendrites receiving impulses from others, and this neural net can make fresh connexions all your life to learn new skills, like surfing.

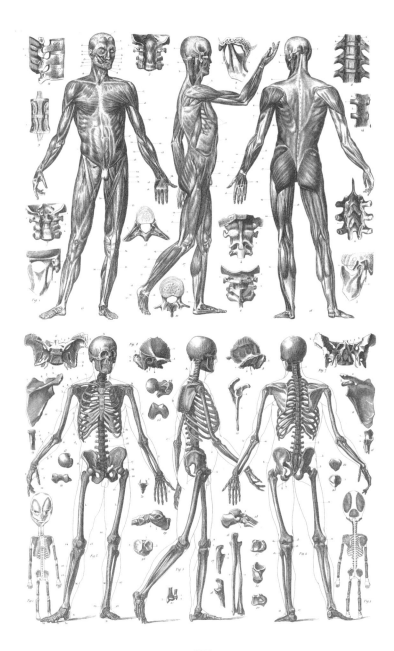

ENDOCRINOLOGY
hormonal biorhythms

Hormones dictate what phase body systems are in, from growth and *metabolism* to immunity and reproduction. They do this by switching influential processes on and off in distant cells. Seven major endocrine glands send hormones, via the blood, to all cells.

Any given hormone has certain 'target' cells that respond to it. Some, like *thyroxin* and *insulin*, target most cell types, in type-variable ways. A few, like growth hormone and its opposite *cortisol*, target all cells. But most endocrine business is self-regulation, via control hormones which only target cells in other glands.

The brain senses and governs the system in the hypothalamus, whose interplay with the pea-sized pituitary, on its stalk, releases control hormones to the lower glands, which make real hormones. Control loops time phases like the morning rush of cortisol, and the monthly surge and ebb of female *sex hormones*. Hormones start phases by switching on master genes that activate groups of other genes, often for enzymes that perform a related set of processes.

Top of the pile, deep in the midbrain, is the mysterious pineal, primal neuro-endocrine uberface. It times major phase changes, like night and day, seasons, puberty, menopause, and maybe death.

Indians have said for years that the simplest way to describe anyone's current state of health is as the activity and interplay of the seven major chakras that sit up and down the midline (*opposite*).

Neuro-endocrine overlap is widespread, and it's hard to tell what's a hormone (*see paracrine*). The adrenal *medulla* is ectodermal (*see page 282*), so is specialized nerve, which fits with that *adrenaline* feeling.

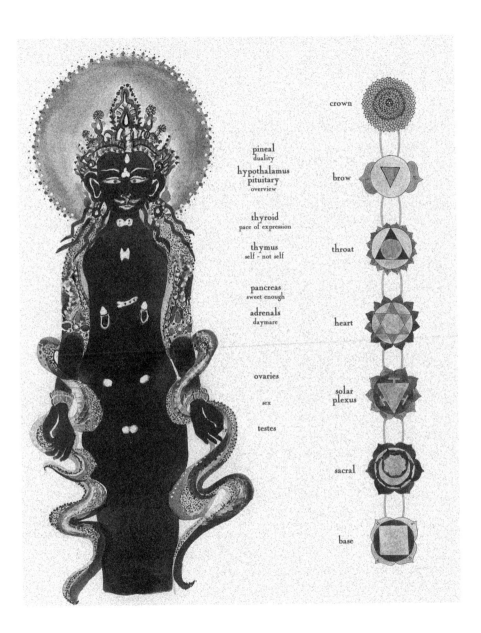

pineal
duality
hypothalamus
pituitary
overview

brow

crown

thyroid
pace of expression

thymus
self - not self

throat

pancreas
sweet enough

adrenals
daymare

heart

ovaries

sex

testes

solar
plexus

sacral

base

HOMEOSTASIS
patterns in status quo

Imagine all the various communications going on at this very moment between all the people on this planet, then cram all that into the same size as you, and that's about how busy it is in there.

This subconscious frenzy has one goal—to maintain equilibrium. Consider circulating calcium levels, honed by *parathyroid hormone* (PTH) and *calcitonin*. High calcium stimulates calcitonin release, to which signal bone cells respond by crystallizing calcium, while kidney cells excrete it, reducing current levels. PTH has opposite effects, and raises calcium levels when they fall. Together these two negative feedback loops keep calcium at the desired level.

Sets of similar loops balance everything from levels of ions to blood pressure, temperature, posture, vision, sanity, and so on, using chains of electrical and chemical signals. Signalling between cells creates integrated patterns of activity, by turning processes on and off inside cells using secondary intracellular signals. Gene switching, usually under hormonal control, tends to govern the longer term variations in cellular activities, whereas short term changes are mainly under electrical control by nerve impulses. Neuro-endocrine integration of these regulatory feedback loops keeps the body's patterns in *status quo*, which is 'homeostasis'.

Positive feedback loops are rare, as they amplify wildly. Apart from triggering *ovulation*, they occur in communication between humans, helping social equilibrium. They often involve the hormone *oxytocin*, which is busy during childbirth, breastfeeding, and orgasms. Oxytocin also mediates the spine-tingling music reflex, and its limbic system effects say "I love you".

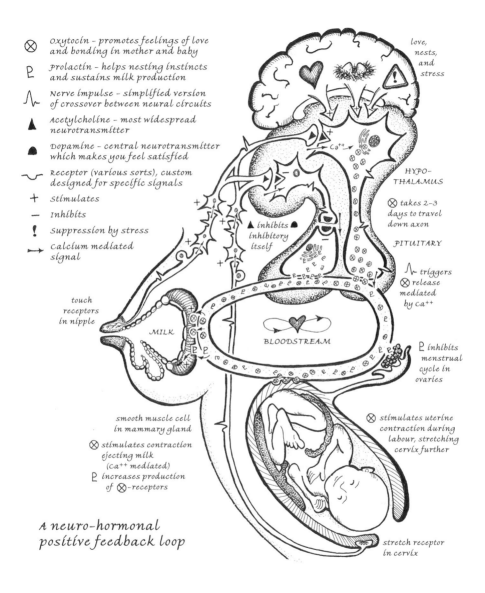

⊗ Oxytocin - promotes feelings of love and bonding in mother and baby

Ρ Prolactin - helps nesting instincts and sustains milk production

Λ Nerve impulse - simplified version of crossover between neural circuits

▲ Acetylcholine - most widespread neurotransmitter

◗ Dopamine - central neurotransmitter which makes you feel satisfied

⌣ Receptor (various sorts), custom designed for specific signals

+ stimulates

− inhibits

! suppression by stress

⟶ calcium mediated signal

love, nests, and stress

← Ca++

HYPO-THALAMUS

⊗ takes 2-3 days to travel down axon

PITUITARY

Λ triggers
⊗ release mediated by Ca++

▲ inhibits ◗ inhibitory itself

touch receptors in nipple

MILK

BLOODSTREAM

Ρ inhibits menstrual cycle in ovaries

⊗ stimulates uterine contraction during labour, stretching cervix further

smooth muscle cell in mammary gland

⊗ stimulates contraction ejecting milk (Ca++ mediated)

Ρ increases production of ⊗-receptors

A neuro-hormonal positive feedback loop

stretch receptor in cervix

311

NIGHT AND DAY

which need each other

The reason you feel as if you're falling apart after you've been up all night is because you are. The hormones that get you going in the morning, like the *catabolic* king, cortisol, make your body break down and use up its resources, to go hunting and nesting. Maintenance and repair work, and *immunication*, are suppressed all day long, so by bedtime, your body is literally disintegrating.

Growth hormone, the *anabolic* queen, rules by night, when your body reunifies the chaos left by the day. For this to happen, communication pathways that are closed by day open up, and like the intense repair of *inflammation*, the body's nightly healing is painful. You can feel where it's going to be busiest as you're dropping off, and, but for sleep, the night would hurt.

Left unfettered, body processes are bursting to go, and have to be firmly regulated if order is to prevail. Patterns of sensation particularly emerge through most nerves being damped. The day hormones' effects on sensory nerves are inhibitory, because all that sensation, while you're actually awake, would hurt too much.

The brainstem gets final short-term say over whether you're asleep, but longer-term patterns of sleep and wake are under the control of the pineal, the original light-sensitive organ. This has evolved inward, but still receives light, via several neural routes, from the eyes. The neuro-endocrine clock in the brain has an innate 25-hour rhythm, but is reset every day to Earth time by the rising Sun, via the pineal. So this gland, which oversees major life changes, is itself ruled by our motion in the solar system.

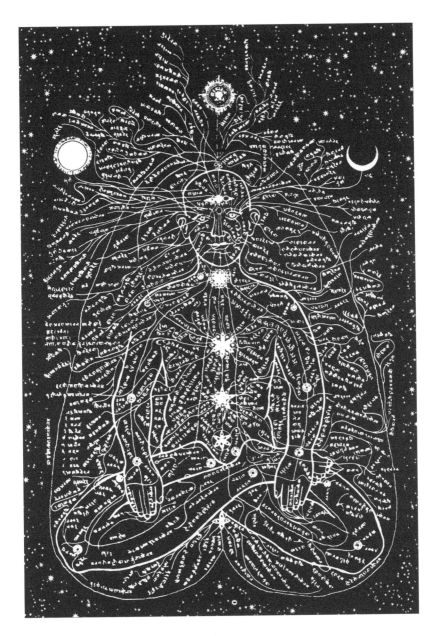

WHERE ARE WE GOING?
what is it for?

The Earth has been revolving around our Sun for five billion years or so, and for most of that time life has been coevolving and refining itself to produce beautiful and complex lifeforms like the human body, which has appeared in the most recent twinkling of the cosmic eye. The entelechy with which evolution realizes new possibilities resembles less a series of lucky breaks than a system that has the idea first, then works out a way to make it happen.

The biggest recent changes in human form are jaw shape and tooth alignment which have altered in only the last few thousand years to cope with cooked food. Other subtler progress may be taking place under our noses, but our bodily structure has been very stable since the image opposite was painted on a cave wall in Spain 8,000 years ago, and for a good 100,000 years or more before that.

Unlike all other organisms, as far as we know, we are in the odd position of having minutely analyzed our inner workings, and we can influence ourselves in novel ways. Biotechnology is now racing ahead of any natural longer-term refinements in our bodies. Many agree that old-style evolution is probably over for humans, and that any major future changes are likely to be self-inflicted. But if mortality isn't a failing—even the stars in the sky die in time—then perhaps the human body is already a perfected form.

Meanwhile, your body is hectically sending signals around itself, generating electrochemical and mechanical patterns, in order to enable you to do things like read this book. Perhaps this continual inner messaging hints at what it's really for, what humans are uniquely good at, which is communication.

315

BOOK VI

10^{10} 10^9 10^8 10^7 10^6

10^5

10^4

1000

100

10

1 light year

10^3

10^4

10^5

100

10

1 AU

0.1

10^6

10^5

10^4 miles

The visible universe, from the Earth's core to the horizon,
14 billion light-years away.

THE COMPACT
COSMOS

Matt Tweed

INTRODUCTION

Gazing out into the vastness of space a complex and intricate tapestry is being revealed by modern science. The essential paradox of existence has inspired generations of wonderers to look to the skies and beyond, and even now we are barely dipping our toes into the vastness of space.

This little book explores some of the ideas painstakingly put together by many great minds attempting to understand and appreciate this magnificence. In the pages which follow distances are given in miles and light-years. Light is the fastest thing there is, travelling 186,000 miles every second. At this speed we could fly round the Earth in a seventh of a second, reach the Moon in one and one third seconds, the Sun in 8 minutes, and the furthest planet in the solar system in a little over four hours. In a year, light covers just under six trillion miles, one light-year, a quarter of the way to Proxima Centauri, our neighbouring star, and then in galactic terms, we've only popped next door.

At a point in time where humanity is waking up to the enormous changes happening on our own tiny planet, the heavens are revealing spectacular scenes of star-birth and calamity, splendor on a scale almost beyond imagining. Long ago, shamans believed that every stone and tree had a guardian spirit, while a modern mystic might say the same of the coiling plasmas that shape nebulæ and galaxies. Our place in space may be one of an infinitude, or as unique as the rarest of gems.

Life, and an awareness to experience it with, is surely a blessing beyond compare.

THE BIG BANG

from small beginnings

Long ago, according to a recent story, seemingly from nowhere, a pinpoint of intensely hot, immensely dense energy appeared. This tiny *singularity* seeded our entire universe. Conditions were so extreme we have only the vaguest idea of the exotic physics at the start of time.

With the young universe cooling fast and space expanding at close to the speed of light, first photons, then quarks and leptons condensed out of the fizzing quantum vacuum like mist on a cold window to form a quark-gluon plasma sea. Next, after one millionth of a second, the quarks combined into hadrons, primarily protons and neutrons, while vast amounts of matter and antimatter wiped each other out leaving only a billionth of the original material, along with vast quantities of gamma rays. Roughly a second after the birth of the universe its temperature dropped enough to crystallize whizzing neutrinos from the photons. Nucleosynthesis started around this time, with protons and neutrons joining to form the nuclei of helium, deuterium, and lithium. Ten minutes later matter consisted simply of three parts hydrogen to one part helium. The universe was expanding incredibly fast, and after two hours there was no longer the density of neutrons to allow any heavier nuclei to f o   r   m   .

When the universe was 377,000 years old it finally became cool enough for electrons to settle into orbits around atomic nuclei, and for the next 100 million years everything remained dark as the vast ionized cloud of hydrogen and helium expanded. Eventually, however, the photons were set free from the plasma, the heavens became transparent, and the infant universe was unveiled in all its newborn glory.

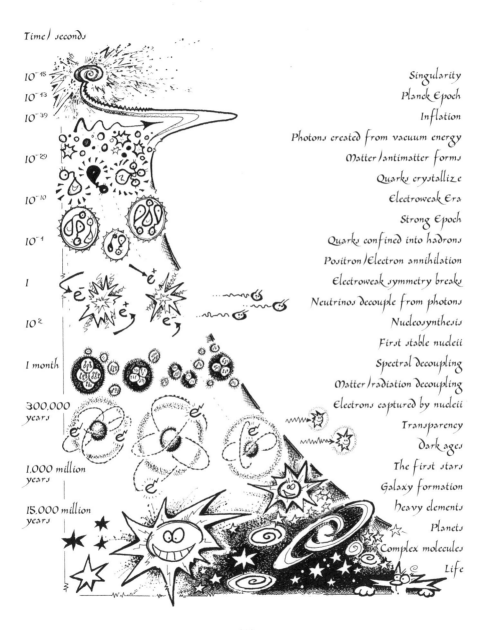

Time / seconds

10^{-45} — Singularity

10^{-43} — Planck Epoch

10^{-39} — Inflation

Photons created from vacuum energy

Matter / antimatter forms

10^{-29} — Quarks crystallize

Electroweak Era

10^{-10} — Strong Epoch

Quarks confined into hadrons

10^{-4} — Positron / Electron annihilation

Electroweak symmetry breaks

1 — Neutrinos decouple from photons

10^{2} — Nucleosynthesis

First stable nucleii

1 month — Spectral decoupling

Matter / radiation decoupling

Electrons captured by nucleii

300,000 years — Transparency

Dark ages

The first stars

1,000 million years — Galaxy formation

heavy elements

15,000 million years — Planets

Complex molecules

Life

THE EARLY UNIVERSE
starbursts at the edge of time

The first bodies to emerge from the chaos were quasars, the most powerful, luminous objects in the universe, early active galaxies built around young supermassive black holes, forming on slight inconsistencies in the otherwise uniform expansion of the universe. Soon afterward, inside and outside of these early *protogalaxies*, a bloom of huge stars burst into life, gorging themselves on primal hydrogen and helium before exploding, seeding newly minted elements into the mix. For the next 500,000 years, until the universe's one billionth birthday, quasars and early stars hatched, lived, and died, recycling earlier generations, and pouring out intense radiation that reionized their surroundings. Ninety-nine percent of all matter in the universe remains in the form of fizzy ionized plasma from this time.

Zoomed out, the universe today looks like a mess (*opposite top left*). Zooming in, however, structures appear, and the final picture opposite shows a supercluster of over 20,000 galaxies. We join our guides here (*below*), at the edge of the visible universe, on their epic journey to the planet we call home. We will encounter sights of amazing beauty and awesome power amongst the starry tapestry. Fasten seat-belts, please.

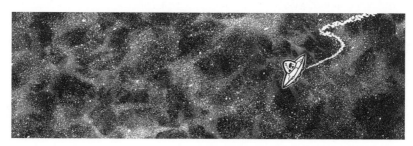

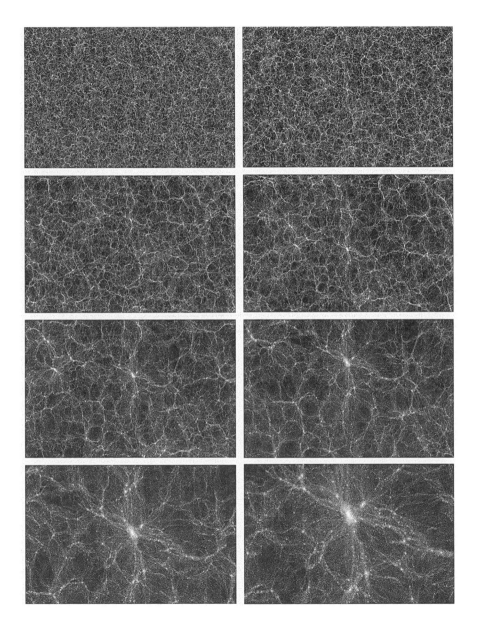

GREAT WALLS AND VOIDS
seriously large-scale structures

In many ways the universe resembles a froth of bubbles. Each cell is a mind-numbing expanse of virtually nothing that stretches for hundreds of millions of light-years. Surprisingly however, the vacuum of deepest space, emptier by far than the best made in labs, still has its own eclectic population. The *intergalactic medium* has the odd lonesome ionized atom floating about, a few for every cubic meter of space, roughly one million hydrogen atoms to each one of the sparser heavier elements.

The biggest voids can't even manage this. A possible Eridanus super-void, covering a dull 500 million light-years, seems responsible for a chilly spot in the near uniform glow of the *cosmic microwave background* radiation, the last echoes of the energetic toddler tantrums of the universe. As deeper patterns fade, new ones appear (*see page 360*).

Around the voids, where two or three bubbles meet, filaments of galaxies hundreds of millions of light-years long braid the edges, entwining like dendrites connecting neurons. As if seeking consolation in numbers, the threads tie together in glowing walls of multi-million strong congregations of galaxies. Estimated at 3.5 billion light-years long by 2.5 billion wide and 50 million thick, the Centaurus Wall (*opposite*) is one of several that form the largest structures so far discovered. Matter this far apart is not yet totally beholden to gravity, continuing instead to flow mainly with the greater cosmic expansion.

At this scale (*opposite*), our travellers can't yet see individual galaxies, they simply appear to twinkle like the squillions of stars they each contain. However, as they continue to fly closer to their destination discrete structures finally begin to appear.

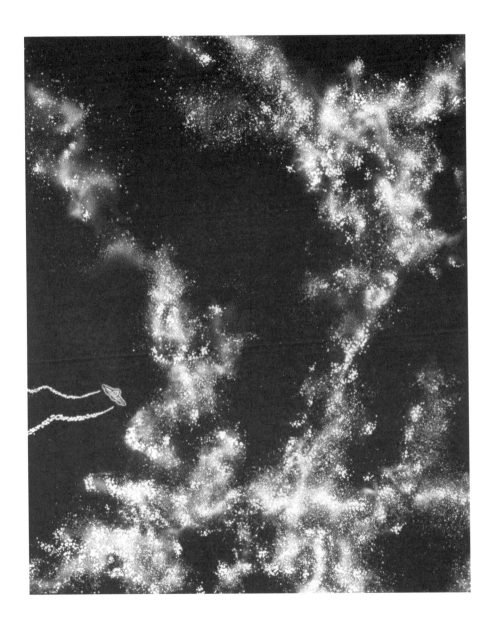

SUPERCLUSTERS
galactic gatherings

Up close, the great glowing walls resolve into smaller chains and sheets of galaxy superclusters, now close enough to experience the pull of gravity. The largest superclusters span hundreds of millions of light-years and contain many thousands of galaxies. We belong to the Local or Virgo Supercluster, a modest 150 million light-years across, a flock of over a hundred smaller galaxy clusters, all being pulled toward a Great Attractor, seemingly an enormous mass hidden somewhere deep within the Centaurus Wall (*previous page*).

The Local Supercluster is centered around the rich Virgo Cluster, 54 million light-years distant. Presided over by one to three giant elliptical galaxies, *rich clusters* are 10–30 million light-years wide, and can embrace 10,000 galaxies. The Virgo Cluster itself is home to the M87 giant elliptical and an eclectic population of 2,000 smaller galaxies. The huge numbers of galaxies in the Virgo Cluster have quite an effect on the neighbourhood, slowing down surrounding poor clusters and pulling them in to make itself even bigger and more attractive. *Poor clusters* are more common than rich ones, with tens of members, mostly spirals, and sizes up to 3 million light-years across. Our Local Group is a poor cluster with two large spiral galaxies and thirty or so others of various sizes, lodged halfway along a filament strung between the Virgo and Fornax clusters (*see maps pages 400-401*), being pulled inexorably toward the center of the Virgo Cluster.

Our travellers have now reached Fornax. If they looked at Earth through a powerful telescope, they would see dead dinosaurs and some sooty mammals, since light takes 65 million years to bridge the gap.

GALAXIES
whorls and wisps

For ease, galaxies are grouped by shape and features in the Hubble Sequence. Three quarters of those visible are *Spiral galaxies*, flattened disks with a central bulge, some with distinct bars like our own (*see SBb below*). Younger stars are found in the disk whilst older stars populate the bulge or surrounding halo. Vaguely spherical *elliptical* galaxies mainly contain older stars, and are thought to be the result of fusions of spiral galaxies, with *lenticular* galaxies the halfway stage. Messy *irregular* galaxies have shapes sculpted by more complex interactions. Surveys suggest that dwarf and low-surface-brightness galaxies, dark and difficult to detect, perhaps outnumber all the rest.

Three of our galactic neighbours are visible to the eye as glowing wisps in the night sky. Two, the large and small Magellanic Clouds, are dwarfs, satellites of our own Milky Way, a mere 170,000 and 190,000 light-years away respectively. The third is our spiral companion, Andromeda, twice our mass. Currently over 2 million light-years away, we are destined to merge in about 5 billion years, ploughing through each other in a swirling maelstrom of arms. Although stars themselves rarely collide, it may be a bit of a rough ride.

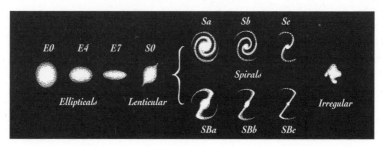

ACTIVE GALAXIES
and quasar questions

With ingenious devices scanning the heavens in spectrums beyond the visible, many marvels have been revealed in recent years, and some of the most awesome are the aptly named *active galaxies*.

Radio galaxies spew jets of matter many light-years long from a relatively small nucleus, and are brightly lit in the x-ray and radio frequencies even though they may be optically dim. *Seyfert galaxies* are spinning discs with intensely active highly ionized nuclei. Elliptical *N-galaxies* have nuclei that vary in intensity, and a subspecies, *BL Lacertæ* (*BL Lac*) *objects*, have active nuclei that are even more wildly fluctuating, probably as a result of looking head-on into a jet. *Quasars* (*quasistellar radio sources*) are a hundred to a thousand times brighter than normal galaxies and emit vast quantities of radiation (*opposite top*). It is thought that they are the bright hearts of active galaxies, powered by supermassive black holes. With very high redshifts suggesting very great distance, quasars are thought to be young galaxies from the earlier ages of the universe.

However, there are puzzles. Twin quasars have been observed each side of younger, lower redshift active galaxies, suggesting that they might be connected, hence at the same distance (*lower opposite*). This raises the possibility that redshifts are not always caused by recession speed and so might not be such reliable measures of distance, although the effect could be caused by gravitational lensing (*see page 363*). Additionally, some jets have extreme redshifts, appearing to be moving faster than light, breaking the cosmic speed limit. These seemingly super-luminal emissions may be due to serious spacetime distortions in the high-energy zones dominated by the huge black holes from which they have burst.

X-ray contour maps of the Virgo Cluster (left) and the Seyfert galaxy
Markarian 205 (right) showing apparent bridges between supposedly distant
high red-shift quasars (QSO) and lower red-shift 'foreground' objects.

BLACK HOLES
through the event horizon

Black holes are the most enigmatic objects in the universe. Put too much matter in one place and immense attraction eventually forces atoms to totally degenerate, crunching neighbouring nuclei together and collapsing electron shells completely. The result is a tiny yet stupendously massive object with gravitational fields so strong that anything nearby, including passing light, is sucked in forever. Just a teaspoon of black hole material would sink to the Earth's core and then suck in the planet.

Supermassive black holes of 10 to 100 billion solar masses probably lurk at the cores of active galaxies and power the fantastic outpourings of quasars. *Massive black holes* are thought to anchor most galaxies. Millions of times the Sun's mass, they coevolve with their galaxy, one shaping the other. *Stellar-class black holes* meanwhile are mere lightweights of 3 to 14 times the Sun's mass, the collapsed remains of large stars.

Those foolhardy enough to approach a black hole (*fig 1.*) would first notice Einstein rings, as light from background stars is lensed by the increasing curvature of spacetime (*2*). The point of no return is the *event horizon*, where light is pulled into an endlessly orbiting photosphere and all within is forever hidden (*3*). To those outside, the traveller seems to slow to a stop at the horizon's edge, frozen in time, redshifting, and fading (*4*). The gravity gradient becomes so steep that the unwary undergo 'spaghettification', with head stretched miles from toes (*5*). Inside a black hole time warps into space, spacetime compressing to an infinitely dense, inescapable singularity (*6*). Ripping the fabric of the universe, wormholes possibly funnel through to other universes. After a massive radiation burst at the wormhole's mouth a white hole might come as light relief (*7*).

fig 1.

fig 2.

fig 3.

fig 4.

fig 5.

fig 6.

Non-spinning
black hole

Spinning
black hole

fig 7.

THE LOCAL GALAXY
meeting the milky way

Parts of the Milky Way galaxy formed around 13.6 billion years ago, when the universe was 100 million years old, making it one of the earliest. Approaching from a distance, our travellers first see what appears to be a starry whirlpool, before they next encounter the *halo*, a spherical shell of dust, gas, and clusters of stars deep into intergalactic space. Elegant arms trace a logarithmic spiral, roughly 100,000 light-years across and 1,000 thick, believed to be shaped by spreading density waves. Between the arms lie dark, dusty lanes. Our own solar system is quietly tucked away in the Orion Spur (*arrowed*), roughly two thirds of the way out.

Joining the galactic arms together is a vast bar, 27,000 light-years long, with a central bulge. At the very center, anchoring it all, lurks *Sagittarius A**, thought to be a monstrous spinning black hole of over 4 million solar masses. Surrounding this are thread-like plasma structures and supernova remnants (*labelled SNR below*), all rapidly orbiting at a slight tilt to the main galactic disk along with an accompanying bunch of neutron stars and smaller black holes.

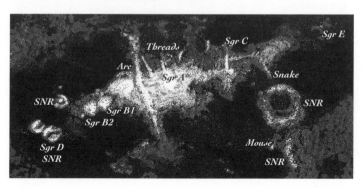

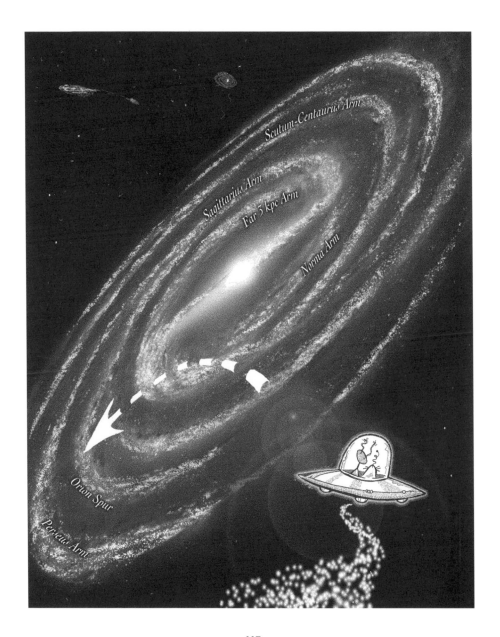

GLOBULAR CLUSTERS
old stars and young stars

The stellar inhabitants of galaxies like ours are divided into two kinds, old (*population II*) and younger (*population I*) stars.

Born in the earliest epoch of the galaxy, and with some estimated to have been around for 13.3 billion years, the older stars are mainly cool red giants, found either orbiting in the central bulge or grouped together in beautiful wandering *globular clusters* (*opposite top right*). Like fairy lights in the void, with eccentric orbital paths taking them far from the galactic plane, globular clusters hold tens of thousands to several million stars. Cradled by mutual gravity, and possibly centered on intermediate-mass black holes, these mature congregations measure from 60 to 300 light-years across. Their leisurely circuits last millions of years, swooping up and out 130,000 light-years into the halo before plunging back through the disk (*opposite top left*). Most suitably large galaxies seem to have their own flock of widely orbiting globular clusters.

Meanwhile, young population I stars like our own Sun are found either alone or mingling in loose open clusters in the gassy, dusty nursery regions of the spiral arms (*lower opposite*). These stars have a higher heavy element content than their population II cousins, having been seeded from the salvaged remains of older stars. With near-circular orbits about the galactic hub, young stars happily sit within the 300 light-year thickness of the galactic disk.

A first generation of live fast, die young, *population III* stars are thought to have graced the primeval universe. Colossal hydrogen and helium monsters a hundred times bigger than the Sun, their short lives would have forged heavier elements for the following generations.

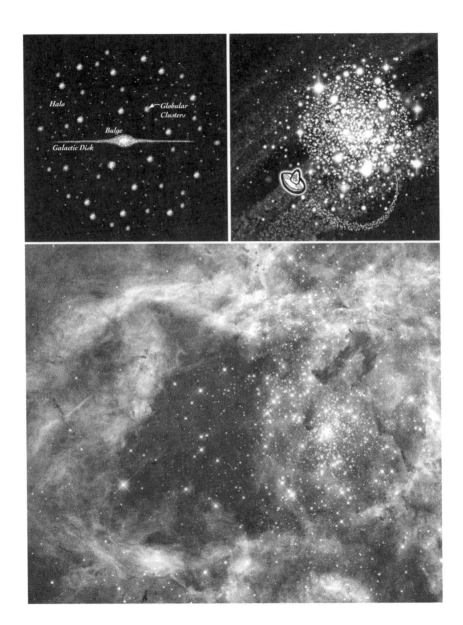

THE MAIN SEQUENCE
growing up and getting old

We are born of stars, and all our atoms, except for the primal hydrogen and helium, are created in these bright furnaces of space. Having first slowly coalesced from clouds of interstellar gas, stars then evolve in a number of ways, depending on their mass, in a potent balancing act between gravity pulling inward and powerful thermonuclear reactions pushing outward (*opposite top*).

Protostars of less than a tenth of a solar mass skulk about as *brown dwarfs*, which never ignite. For newly born protostars with only a third of the Sun's mass, gravity squeezes the core to a temperature high enough for nuclear fusion to commence. Four hydrogen nuclei are squeezed together to make one very slightly lighter helium nucleus, releasing the excess mass-energy as positrons, neutrinos, and a burst of energy—starlight. These *red dwarfs* use their fuel frugally, shining for hundreds of billions of years before fizzling out into a *black dwarf*.

Mid-mass *main sequence* stars like our own Sun take a different path. Hydrogen fusion carries on for 10 billion years or so until the hydrogen fuel runs out and gravity takes over. The enormous pressures generated as the star collapses in on itself cause soaring temperatures, fusing helium into new elements (carbon, oxygen, and nitrogen), puffing the star off of the main-sequence and into a much larger *red giant*. After a few billion years when even this fuel is finished, the core collapses again, but lacking enough mass to generate sufficient heat to restart fusion the exhausted star blows off its gassy outer envelope into a nebula and enters long old age as a hot, compact *white dwarf*.

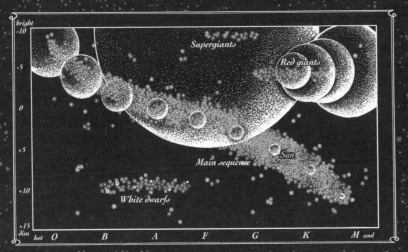

The Hertzprung-Russell diagram plots stars according to their luminosity, expressed as absolute magnitude (vertical scale), and spectral class (horizontal scale). Running diagonally across the diagram is the main sequence where most stars spend the greater part of their lives.

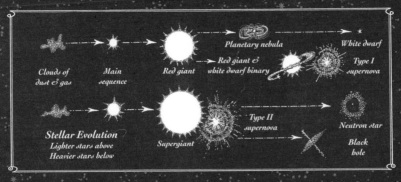

The eventual fate of a star depends on its mass, the larger the star the more spectacular its demise. Small stars fade away into a quiet old age, whilst larger stars spectacularly explode as supernovæ then undergo immense gravitational collapse into ultra-dense neutron stars or inescapable black holes.

ELEMENTAL FACTORIES
forges of the heavens

Stars above eight solar masses lead frantic lives relative to their smaller cousins, growing through several stages as successive fusion processes convert their changing fuel into heavier and heavier elements, swelling them into *supergiants*, hundreds of times the sun's diameter.

Racing through the hydrogen and helium phases, these stars' much greater mass ignites carbon fusion, scrunching nuclei together to form neon, oxygen, and magnesium. As the carbon is exhausted the core contracts again, this time cooking up silicon and sulfur. Intermediate elements are meanwhile painstakingly assembled in the core by *slow neutron capture*, like an intricate jigsaw that takes thousands of years.

With a further collapse the core is converted into cobalt, nickel, and finally iron, the hinge of the universe, when fusion stalls (all elements up to iron in the periodic table give out energy as they are fused in stars, while elements after iron need energy to form). At this point gravity wins once more and with a final core collapse the star undergoes a drastic final implosion that ends in a explosive supernova. In this immense cauldron many new elements are concocted. Some are created by *rapid neutron capture* with nuclei being slammed together to form heavy atoms like uranium, whilst others are formed by *spallation*, where larger atoms are chipped into smaller pieces by ultra high speed debris as they whizz through space (*opposite*).

The eventual fate of large stars depends on their mass, either folding in to become a dense degenerate neutron star, or suffering total gravitational collapse to become an inescapable black hole.

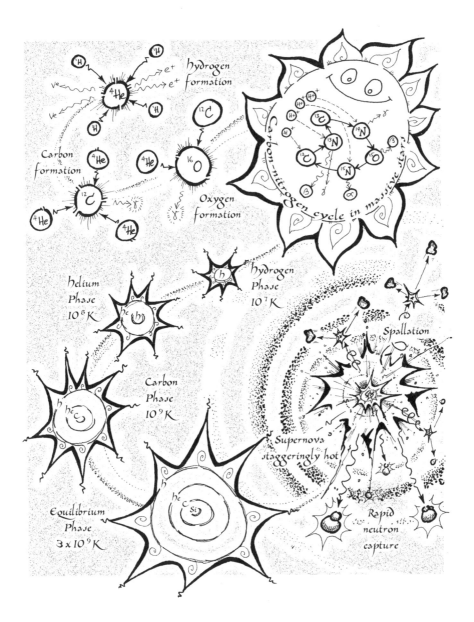

hydrogen
formation

^1H ^1H ^1H
$\rightarrow e^+$
$\rightarrow e^+$
^4He
ν_e
ν_e
^1H ^1H

Carbon-nitrogen cycle in massive stars

$H+$ $H+$
^{13}C ^{14}N γ
$H+$ ^{13}N ^{15}O
^{12}C ^{15}N e^+

Carbon
formation

^4He ^4He ^{12}C ^{16}O
γ

Oxygen
formation
γ

^4He ^{12}C ^4He
^4He

helium
Phase
$10^8 K$
he (h)

hydrogen
Phase
$10^7 K$
h

Spallation

Carbon
Phase
$10^9 K$
h he c

Supernova
staggeringly hot

Equilibrium
Phase
$3 \times 10^9 K$
h he c Si

Rapid
neutron
capture

Neutron Stars and Novæ
lighthouses and candles

The dazzling supernova deaths of four to eight solar-mass stars (*below*) leave behind exotic iron-shelled *neutron stars*. Gravity, no longer balanced by fusion, crushes the remains of the core. Atoms inside are scrunched so tightly that the electrons combine with the protons, leaving nothing but neutrons. Neutron stars are incredibly compact, a mere ten miles or so across, and super dense, a thimbleful on Earth would weigh as much as a mountain. Such high densities lead to extraordinary gravitational and magnetic fields. Spinning neutron stars create jets of energized matter that emit intense beams of light and radio waves. When a beam sweeps the Earth, the star pulsates like a celestial lighthouse. The fastest *pulsars* blink nearly a thousand times a second (*opposite top*).

Useful for astronomers, *Cephid variables* are stars whose size and luminosity vary with clockwork regularity. Cephids with the same period are equally bright, so dimmer ones are further away. This means they can be employed as standard candles, enabling the distance to other stars in their vicinity to be estimated (*center opposite*).

About a third of a galaxy's stars dwell in binary, triple, or even larger multiple systems. Binaries co-orbiting at a distance may quietly lead separate lives. However, in closer binary systems the more massive star can start borrowing stuff like a bad neighbour (*lower opposite*).

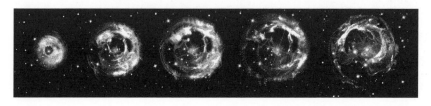

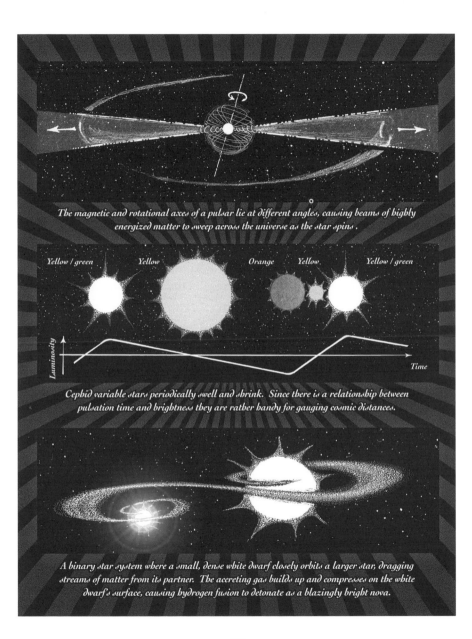

The magnetic and rotational axes of a pulsar lie at different angles, causing beams of highly energized matter to sweep across the universe as the star spins.

Yellow / green   Yellow   Orange   Yellow   Yellow / green

Luminosity

Time

Cephid variable stars periodically swell and shrink. Since there is a relationship between pulsation time and brightness they are rather handy for gauging cosmic distances.

A binary star system where a small, dense white dwarf closely orbits a larger star, dragging streams of matter from its partner. The accreting gas builds up and compresses on the white dwarf's surface, causing hydrogen fusion to detonate as a blazingly bright nova.

Nurseries and Nebulæ
stellar birth and death

Heralding the beginning and end of a star, *nebulæ* are truly spectacular sights. Giant molecular clouds of hydrogen, helium, and a few heavier elements, as well as ammonia, water, and carbon dioxide, form the *diffuse nebulæ* delivery rooms for stellar nurseries (*below*), aglow with the light of new stars, born of gravitation's incessant pull.

Young hot stars excite the blushing plasma of *emission nebulæ*, their ultraviolet light stripping electrons from nearby hydrogen which glows red when the electrons recombine. Add helium, oxygen, and nitrogen to the mix and things get more colourful, scintillating blue and green.

Vast dusty clouds scattering the light of nearby stars cause *reflection nebulæ*, the Pleiades hosting a perfect example of a fetching blue one.

Planetary nebulæ (*opposite*) occur at the end of a giant star's life as the outer layers of the star are blown off into new larger shells of plasma and dust about the exposed core. Layers of ionized gas stream away at different speeds to follow twisting magnetic fields, whilst shock waves jostle the interstellar medium into birthing the next generation.

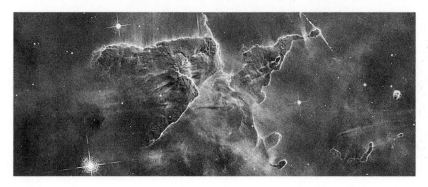

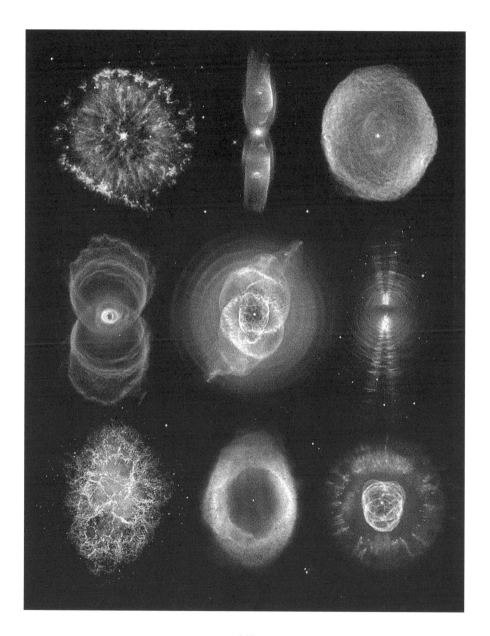

347

AURIC OBJECTS
to the heliopause and beyond

Three quarters of the way from our closest stellar neighbour, *Proxima Centauri*, our travellers encounter the vast *Oort cloud*, thought to be a shell of icy debris left over from the solar system's birth. Flung into highly eccentric orbits a light-year from the Sun, some of these dirty snowballs pay rare flying visits as long period comets. More regularly, short period ones, like Halley's comet, fall in from the distant *scattered disk* and *Kuiper belt*, which extend out 5 billion miles from the Sun. This is also home to the dwarf planets Pluto, Eris, and Makemake.

Further in are four giant planets and their varied rings. Wayfarers amble here too, billions of small rock, ice, and carbon planetoids. Centaurs speckle the gap between gas giants, the inner of which has its own geometrically orbiting camps of Greeks and Trojans marching in front and behind. Partway between here and the four rocky inner planets, multitudes of asteroids wander around their circuits.

From the Sun at the center comes the solar wind, a wild rush of electrically charged particles blowing into space to form the *heliosphere*, a huge magnetized bubble of plasma. Slowed to the speed of sound about 10 billion miles out at the *termination shock*, the ions eventually succumb to galactic influences at the *heliopause*, 14 billion miles away. The last ripples of the Sun's magnetic grip fade at around 21 billion miles out, the bow shock wave of our solar system as it ploughs its way on through the tenuous interstellar medium.

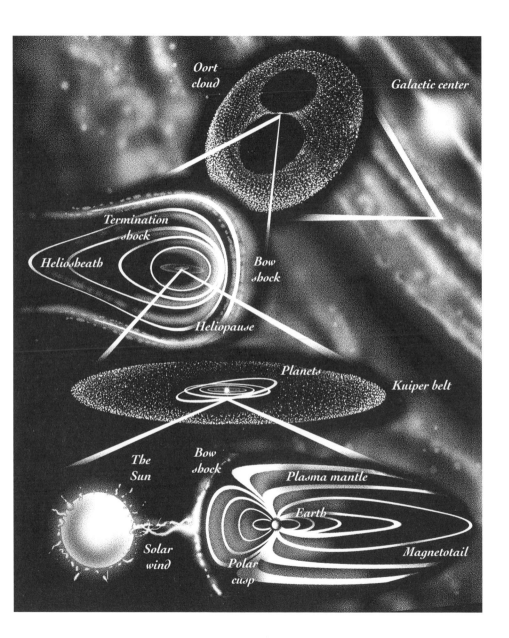

Oort cloud

Galactic center

Termination shock

Heliosheath

Bow shock

Heliopause

Planets

Kuiper belt

The Sun

Bow shock

Plasma mantle

Earth

Solar wind

Polar cusp

Magnetotail

349

OUR NEIGHBOURHOOD STAR
a glimpse of the Sun

At the heart of the solar system, and anchoring the planets, our Sun is a prodigious nuclear factory giving off heat, light, and other radiation. It can take hundreds of thousands of years for a photon of light to barge its way out of the seething interior, and then a fleeting eight minutes to travel the 93 million miles to our travellers' final destination, Earth.

At about 4.6 billion years old, the Sun is nearly halfway through its life. Its present diameter of 870,000 miles will eventually bloat out into a red giant, frying us all and swallowing the Earth before the Sun shrugs off its outer layers into a planetary nebula and retires as a white dwarf.

The Sun is mostly superheated plasma (*see page 364*), 5,800°K at its surface and 15 million degrees at its center. Convection bubbles thousands of miles wide boil across the surface while intensely twining magnetic fields create majestic loops and prominences. Because the Sun's equator rotates faster than its poles, its primary field winds up and then reverses every 11 years, producing sunspots, world-sized windows into the marginally cooler inner layers (*below*).

Periodically, billions of tons of plasma burst forth in titanic *coronal mass ejections*, roiling clouds up to 10 million miles wide ripping through space at 5 million miles an hour. Surges in solar weather can disrupt life for inhabitants of the solar system, messing with their electronics, causing power surges, and knocking out satellites. They also produce spectacular aurora displays at the poles of many planets.

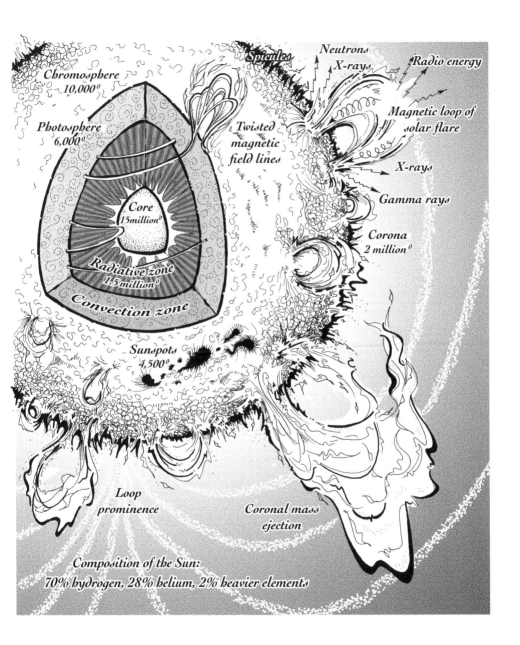

Chromosphere
10,000°

Spicules

Neutrons
X-rays

Radio energy

Photosphere
6,000°

Twisted
magnetic
field lines

Magnetic loop of
solar flare

X-rays

Core
15 million°

Gamma rays

Corona
2 million°

Radiative zone
1.5 million°

Convection zone

Sunspots
4,500°

Loop
prominence

Coronal mass
ejection

Composition of the Sun:
70% hydrogen, 28% helium, 2% heavier elements

THE SOLAR SYSTEM
formation of planets and moons

The solar system began to form about 4.6 billion years ago when, disturbed by a nearby supernova, a huge cloud of dust and gas began to collapse, and over the next 100,000 years rotated and flattened into a disk. At the center a lump became so massively squeezed it ignited as the Sun.

Over the next ten million years gases in the disk were blown to the outer regions where gas giants began to form. It then took another 100 million years for smaller planetesimals orbiting in the innermost rocky region of the accretion disk to coagulate into the embryonic inner planets (*below*). In the final 100,000 years, like a mixer on full, some ferociously collided whilst others were flung out into space. The young gas giants gobbled up much of the remaining matter, clearing their orbits. Moons were either captured, or formed through violent collisions. A powerful unstable 2:1 resonance between Jupiter and Saturn, then orbiting closer to the Sun, caused Uranus and Neptune to migrate outward into the Kuiper belt, precipitating the late heavy bombardment which hurled comets and meteorites all over the place, causing craters on many young worlds.

Luckily for us, after a billion years of turbulent evolution, the solar system settled down a little.

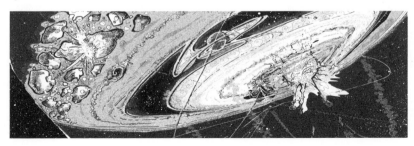

MERCURY
small rocky world with no moons, each day is two years

VENUS
hot high pressure sulfuric acid world with no moons

EARTH
wet atmospheric world with large moon the size of Mercury

MARS
smallish rocky red world with ice caps and two tiny moons

JUPITER
enormous gas giant with 4 monster moons and 59 smaller ones

SATURN
huge gas giant with rings, one massive moon and 61 smaller

URANUS
large ice giant with very thin rings and 27 smallish moons

NEPTUNE
ice giant farthest from the sun, with 13 small moons

Sun

Greeks & Trojans

Mercury Earth
Venus Mars Asteroid Belt Jupiter Saturn Centaurs
Neptune
Uranus Pluto Kuiper Belt Oort Cloud

0 AU	1		2	4	8	16	52	64	128	256
0 million miles	100		250	500	1,000	2,500	5,000		10,000	25,000

GAS, LIQUID, ICE, AND ROCK
fields, seas, and surfaces

Weather and geology vary massively throughout the solar system. Furthest out, the *ice giants* Neptune and Uranus are mostly comprised of hydrogen and helium, but also have methane and water. Both display curiously tilted asymmetrical magnetic fields, perhaps due to the rotations of their highly conductive superheated liquid interiors. The two largest planets, the *gas giants* Saturn (*lower opposite*) and Jupiter, are also mainly hydrogen and a quarter helium, the same mix as the early universe. Both have thick atmospheres of swirling bands of ammonia and water vapour which smoothly transform to supercritical fluid interiors as pressures and temperatures rise. Still deeper inside, the hydrogen is squished into its liquid metallic form, a shield around their innermost rocky cores.

With dense mostly iron cores and rocky silicate mantles, the four terrestrial inner planets, Mars, Earth, Venus, and Mercury all sport solid crusts. Their atmospheres vary, from the barest wisps on broiling hot Mercury, through to Venus' dense greenhouse of carbon dioxide and hydrogen sulfide. Venus is the most volcanic planet in the solar system, and Mars too once had extensive volcanism and water flowing in its canyons. Today, however, Mars is arid except for its polar ice-caps.

Moons in the solar system also display amazing variety. Neptune's big moon Triton has six-mile-high nitrogen geysers, whilst Jupiter's large innermost moon Io is the most geologically active object in the solar system, its insides churned up by its neighbours. Europa may have warm water oceans underneath its icy surface, and Saturn's massive moon Titan may have life-filled subsurface oceans too.

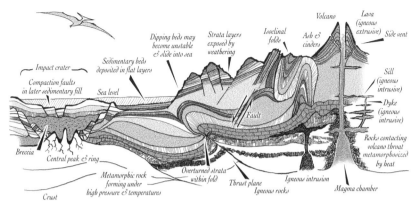

Terrestrial geology in a nutshell. Rocks are made of minerals, and come in three kinds: Igneous Rocks form as molten rocks solidify; Metamorphic Rocks form via heat, pressure, and chemical processes deep in the Earth; Sedimentary Rocks are formed by the accumulation of sediments (including organic matter, e.g. seashells). Folding, fracturing, crunching, etc. of the Earth's crust then mix it all up further. See page 402.

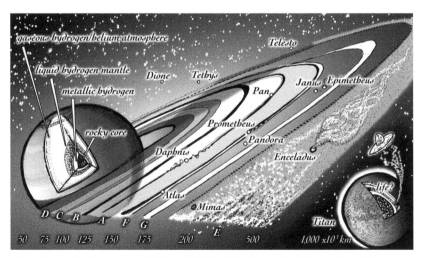

Saturn's amazing ring system, sculpted by gravity's tides. The rings are formed from millions of icy clusters with gaps resonantly shepherded by more than 50 moons . They span 300,000 miles and yet are mostly just 10 yards thick. The largest moon, Titan, has seasons, hydrocarbon seas, and possibly subsurface water oceans. Enceladus spouts geysers 100 miles high from its south pole to form the E ring, Daphnis raises ring waves nearly a mile high, whilst Epithemus and Janus share similar orbits and swap places when they meet.

THIS ISLAND EARTH
the planet we live on

Third planet from the Sun, Earth began its life 4.5 billion years ago as a seething ball of molten lava and raining meteors. Iron and nickel sank to the core, settling through the silicate mantle of fluid rock. The surface had just cooled enough to form a crust when a colliding planetoid melted everything again and knocked off a lump to form our large moon.

Water is thought to have been delivered to Earth from outer colder regions of the solar system, including comets, early in its formation, and the early oceans soon condensed from the cooling CO_2, H_2S, and NH_3-rich atmosphere to cover 71% of its surface. A complex pattern of tectonic plates, winds, currents, and tides developed, cycling water, minerals, energy, and continents around the globe (*opposite top*). The Earth's magnetic field was semi-established a billion years after its formation, a useful shield for the young atmosphere against the solar winds which were then 100 times stronger than they are today. Evolving life changed the atmosphere to today's mix of 78% N, 21% O, 0.9% Ar, 0.04% CO_2, small wafts of other gases, and up to 1% water vapour.

The continents of the Earth are essentially huge and ancient rocky islands which float about on the planetary surface. Between them, and being continuously created from mid-oceanic ridges and destroyed by subduction, is a thin skin of new rock—the ocean floor (*opposite lower right*). In various epochs the islands form vast supercontinents (*below*).

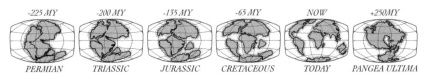

-225 MY   *-200 MY*   *-135 MY*   *-65 MY*   *NOW*   *+250MY*

PERMIAN   *TRIASSIC*   *JURASSIC*   *CRETACEOUS*   *TODAY*   *PANGEA ULTIMA*

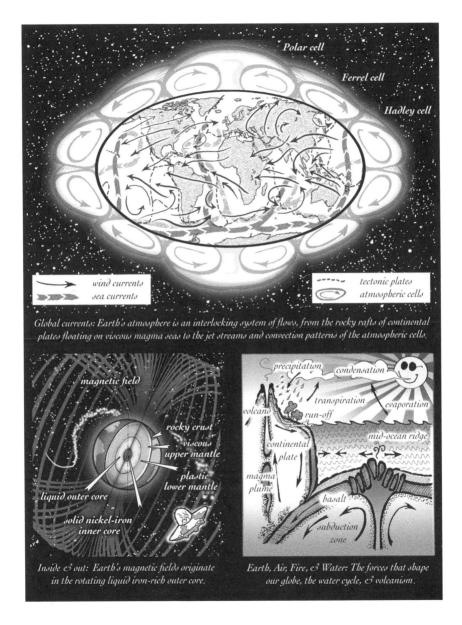

Polar cell

Ferrel cell

Hadley cell

→▶—▶—▶ *wind currents*
⇒⇒⇒ *sea currents*

- - - - - *tectonic plates*
⬭ *atmospheric cells*

Global currents: Earth's atmosphere is an interlocking system of flows, from the rocky rafts of continental plates floating on viscous magma seas to the jet streams and convection patterns of the atmospheric cells.

magnetic field

rocky crust
viscous upper mantle

plastic lower mantle

liquid outer core

solid nickel-iron inner core

precipitation   condensation

transpiration   evaporation

volcano   run-off

continental plate   mid-ocean ridge

magma plume   basalt

subduction zone

Inside & out: Earth's magnetic fields originate in the rotating liquid iron-rich outer core.

Earth, Air, Fire, & Water: The forces that shape our globe, the water cycle, & volcanism.

357

SHIFTING SPECTRA
expanding space

When Earth's scientists began to look out at distant nebulæ, galaxies, and stars, they found that they could tell what they were made of by looking at the light. *Spectroscopy* uses the unique fingerprint frequencies at which atoms and molecules emit or absorb light. Splitting the light received from an object into a spectrum reveals these frequencies as spectral lines, which can then be analyzed to determine the chemical ingredients.

Light can also be used to estimate speed. The *Döppler effect* makes approaching objects appear blueshifted as light waves bunch up toward the blue end of the spectrum, and conversely receding objects appear red-shifted as the light waves are stretched. A complex relativistic version of this idea is used to judge galactic velocities by measuring how far the distinctive patterns of spectral lines have moved along the spectrum. Observations show that faint distant galaxies become increasingly red-shifted, suggesting that the farther away an object is, the faster it is racing away from us. The fascinating implication of this cosmological redshift is that the whole universe is expanding ever faster.

Hubble's Law charts a relationship between an object's brightness and its redshift, to calculate the immense distances involved. The highest redshifts yet found belong to galaxies whose light has taken over 13 billion years to reach us, suggesting we are seeing them as they were in the very early universe, just 600 million years after the Big Bang.

nearby star   far star   close galaxy   distant quasar

Increasing recessional velocity

The spectrum of a hot main sequence star has mainly hydrogen & helium absorption lines.

Cooler red giants have a more complex spectrum showing that heavier elements are present.

Blueshift   Redshift

Doppler shifts: an approaching object appears blue-shifted while a receding one will seem red-shifted.

Increasing redshift

Cosmological redshift indicates that the universe is expanding. Like dots painted on a balloon (or alien) which is then inflated, distant points seem to race away faster than closer ones. Observations show the spectra of distant objects have increasing redshifts.

359

THE RADIATION COSMOS
hidden skies for electronic eyes

Our limited eyes recognize a relatively narrow band of electromagnetic frequencies which we call visible light. Modern instruments, however, can see far beyond these frequencies to reveal amazing sights.

Radio waves have the longest wavelength. Bright in this band are the spectacular jets that stretch for thousands of light-years from the fervid cores of active galaxies. Vast glowing molecular clouds of ionized gas swathe the sky, while the radiant remains of supernovæ allow peeks into the mælstrom that is our galaxy's heart. Raising the vibration a notch, everything is bathed in a near uniform cosmic *microwave* background radiation, thought to be a relic of the super-energetic earliest epochs of the universe. Small ripples in this subtle sea may hold clues as to how today's large structures evolved. Syncopated microwaves also splash from maser fountains around new stars. At higher frequencies stars begin to shine in the *infrared*; from the youngest, coddled in dusty nebulæ, through bright starburst galaxies birthing new stars, to ageing white dwarfs and giants.

Above the infrared is visible light. The Sun emits more energy in this waveband than any other, which is why our eyes have evolved to see it.

On the other side of the visible lies the *ultraviolet*. Stars burn hottest at the beginning and end of their lives, so here we see young UV galaxies ablaze with newly born hot blue stars and supergiants shining bright. Tenuous galactic plasma shimmers in the shortwave *x-ray* bands, with dazzling outbursts marking extreme events, possibly signalling the gravitational maws of ferocious black holes. At greater energies still, occasional blinding detonations of *gamma ray* bursters flash, perhaps the cataclysmic final roars of supermassive stars in very distant galaxies.

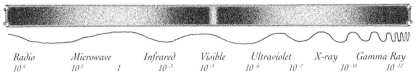

The Electromagnetic Spectrum, wavelengths in meters

Radio	Microwave	Infrared	Visible	Ultraviolet	X-ray	Gamma Ray
10^4	10^2	1	10^{-3}	10^{-5}	10^{-6}	10^{-7}	10^{-10}	10^{-12}

The jets of radio galaxy Cygnus A stretch 18,000 light-years in each direction.

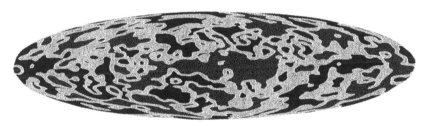

Sky map of variations in the cosmic microwave background radiation. These small fluctuations are believed to have formed the kernels of the later large-scale structure of the universe.

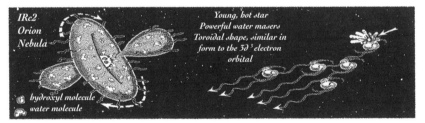

Incredibly strong masers (microwave amplification by stimulated emission of radiation) form around young stars shrouded by molecular clouds. Water and hydroxyl molecules enter excited energy states. Dropping back they emit a buzz that fires up the whole system.

Forces and Fizzicks

the curious case of the neutrino

Holding together everything in the entire universe are just four forces (*shown opposite*). Two, the *strong* and *weak* nuclear forces, govern subatomic interactions, whereas sparks, magnets, aurora, and rays of light are due to the *electromagnetic* force. The weak force and electromagnetism separated from a unified *electroweak* force when the temperature dropped below 10^{15}K, a second or so after the Big Bang.

Important players in the electroweak-dominated ultra-high energy plasmas of the infant universe were *neutrinos*, which decoupled after a second, and ever since have had infrequent weak force interactions with ordinary matter, zipping undisturbed through stars and planets. These tiny neutral lepton cousins to the electron are about two million times less massive, yet still account for one tenth of the known matter in the universe. They come in three 'flavours' of different mass, and unlike the other fermion families (*see page 170*), have the chameleon-like ability to oscillate between flavours as they race along close to lightspeed. As if this isn't enough, they might be their own anti-particle, as only left-handed neutrinos and right-handed anti-neutrinos have been seen.

The last and by far the least of the forces, *gravity* extends its grip across the cosmos by warping spacetime—the greater the mass, the larger the warp (*see page 367*). Both space and time are affected, clocks running slightly slower on Earth than they do in orbit. Huge events like supernovæ or colliding black holes are thought to create *gravity waves*, rippling distortions of spacetime itself. *Gravitational lenses* occur when light from a distant source is bent as it passes a massive intervening object producing spectacular multiple images of the source.

The strong force binds together the quarks that make up atomic nuclei.

The weak force governs radioactive decay & neutrino interactions.

Electromagnetism is not just light & radio waves, but also enables chemistry to take place.

Neutrinos are produced in vast quantities during stellar fusion, though their weak interactions with matter and odd 'flavour' oscillations mean we are rarely aware of them.

Colliding black holes create gravity waves that ripple through space-time.

Light from a distant quasar is bent while passing a massive galaxy, creating multiple images of the source.

A fine gravitational lens centered on Abell 2218, revealing background galaxies.

A PLASMA UNIVERSE

into the vortex

Ninety-nine percent of all known matter in the universe is in the form of plasma, a gaseous, high energy state of stuff where electrons are freed from their atoms. Plasmas conduct electricity extremely well, tightly spiralling electrons through magnetic fields to produce dynamic Birkeland currents which coil in paired filaments. Serpentine flows link Jupiter and its moons (*opposite top*), and filaments over 100 light-years long and 3 light-years wide have been detected in our galactic center. Enormous plasma lobes sit above and below the Milky Way, perhaps the remains of a more active phase of its evolution (*opposite, lower left*).

Plasma ribbons reach far across deep space, and near light-speed jets shoot from the cores of active galaxies. The theoretical maximum size that filaments can achieve before becoming unstable is similar to the largest structures so far observed, the Great Walls. Galactic evolution has been modelled by twisting currents in the laboratory (*below*), and since plasma's behaviour is the same at any scale, they may yet hold the keys to many cosmic riddles, some theorists even claiming that plasma webs on a cosmic scale do away with the need for dark matter.

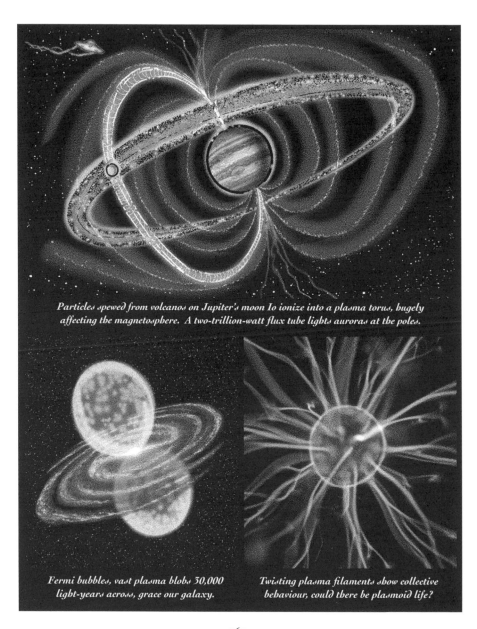

Particles spewed from volcanos on Jupiter's moon Io ionize into a plasma torus, hugely affecting the magnetosphere. A two-trillion-watt flux tube lights auroras at the poles.

Fermi bubbles, vast plasma blobs 30,000 light-years across, grace our galaxy.

Twisting plasma filaments show collective behaviour, could there be plasmoid life?

TIME AND LIGHT BUBBLES
as far as we can see

Our horizon, the limit of what is visible to us, is a bubble expanding at the speed of light (about 186,000 miles per second). Light takes time to travel, so every glance peers into the past, and the more distant an object we spy, the further back in time we are seeing—the light we see from far-off quasars left them billions of years ago, back when the universe was a sprightly youth. A light-year is the distance light covers in a year, just under 5.9 trillion miles and a convenient measure for the vastness of space. Because we can only see back 13.7 billion years (the age of the universe), or (due to the expansion of spacetime) about 75 billion light-years, we simply have no idea how big the universe actually is or what exists outside the horizon of the bubble we can see (*opposite top*).

Einstein's theory of relativity links the three dimensions of space and one of time into a single four-dimensional entity, *spacetime*. A point in spacetime is a sphere expanding at the speed of light, much like the wave function of a photon. Any two times and places connected by the speed of light occupy the same point in spacetime. Spacetime can be imagined as a stretchy sheet (*center opposite*), on which objects affect space and time, causing apparent gravitation and acceleration. Massive spinning objects cause *frame-dragging*, pulling the entire sheet around with them.

Everything is moving at the speed of light. We are hurtling through time at light-speed even if sitting reading a book. If we start to move through space, our velocity through time slows to make the combined space and time velocities still equal to the speed of light (*see page 114*). At our usual slow pace the effects are minimal. However, the faster we go, the more noticeable this curious give and take becomes (*lower opposite*).

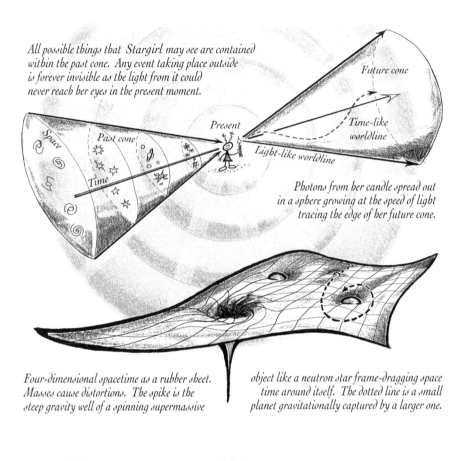

All possible things that *Stargirl* may see are contained within the past cone. Any event taking place outside is forever invisible as the light from it could never reach her eyes in the present moment.

Future cone

Time-like worldline

Space

Past cone

Present

Time

Light-like worldline

Photons from her candle spread out in a sphere growing at the speed of light tracing the edge of her future cone.

Four-dimensional spacetime as a rubber sheet. Masses cause distortions. The spike is the steep gravity well of a spinning supermassive object like a neutron star frame-dragging space time around itself. The dotted line is a small planet gravitationally captured by a larger one.

The alien watches a photon bounce off the UFO roof. Stargirl follows it as he flies by, seeing it apparently go farther.

Light cannot speed up, so it is time that changes, passing more slowly in the spaceship than it is for stationary Stargirl.

SPACETIME GAMES

there and back again

If we travel really fast, relativistic effects become ever more severe as space and time juggle to conserve the speed of light. As an example, let's send an alien on a journey to a distant star, his UFO naturally flying at close to light-speed (*fig.1 opposite*).

One of the first things an observer, Stargirl, notices is that the UFO appears shorter (*2*). This is a side-effect of clocks running at different rates when in relative motion, since the UFO's length from Stargirl's perspective is measured by its speed multiplied by the time taken to go past a point. For the alien aboard, however, his craft's length remains the same.

As the UFO approaches light-speed, more and more energy is needed to accelerate, since rather than boosting the velocity some of the energy is instead converted into mass (*3*). This is a result of Einstein's $E = mc^2$ formula, which says that mass can be changed into an incredible amount of energy, and vice versa. In this instance the quicker the UFO travels the more total energy it has, thereby making it more massive and ever harder to accelerate. If it reached the speed of light its mass would shoot up to infinity, leaving the unfortunate alien undergoing gravitational collapse and ending up as a black hole.

Luckily our alien is wise, keeps his speed down slightly, and manages the round trip to a nearby star in what seems a few months. Alas however, time has passed more slowly for the fast-moving traveller than for those left behind (*4*). On his return, all that remains of Stargirl are her great great great great grandchildren (*5*), who grew up with an old family story of a friend who flew off to a distant star.

fig. 1

fig. 2

$$E=mc^2$$

where E is energy
m is mass
c is the speed of light

fig. 3

fig. 4

fig. 5

369

BIG BANG OR STEADY STATE?
fabricating a universe

Although the *Big Bang* theory is our most popular scientific creation story, there are other takes. One is *steady state cosmology* which leans against a Big Bang and towards continuous creation in a universe that has always existed. Here hydrogen spontaneously forms in interstellar space or galactic cores, slowly building into higher order objects. A *quasi-steady state* variant pictures the universe as expanding and contracting over immense periods, each time building on structures remaining from previous ones.

Cosmic breathing is echoed by the *cyclic braneworlds* model in which our universe is envisioned as a huge four-dimensional bubble, or *brane,* coexisting with others floating about in a higher-dimensional space. Brane universes periodically collide then bounce apart to initiate wonderful new phases of creation.

Ultimately the fate of the universe rests on its *critical density*, a balancing act between gravity and inflation. An *open universe* has too little mass for gravity to stop expansion. Like an infinite three-dimensional saddle, matter dilutes over time and entropy takes over, leaving nothing but huge black holes and fading photons. Alternatively, in a *closed universe,* with more than enough mass to stop expansion, the universe curls in on itself, reaching a maximum size before shrinking and ending its days in a big crunch. Between these extremes, a *flat universe* exactly balances gravity and inflation, eventually reaching a stable state. Intriguingly, some theoretical finite flat universes have no edges and wrap back on themselves. In these, flying in perfectly straight line in any direction would eventually bring any brave travellers back to where they started.

Like everything else, theories are born, reproduce and evolve.

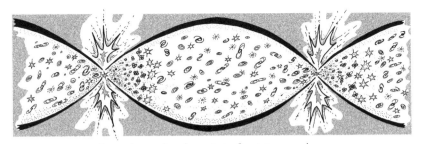

Repeating cycles of quasi-steady state cosmology

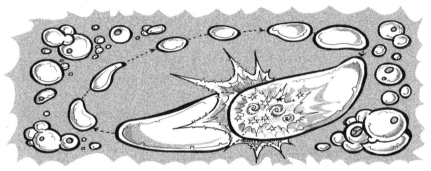

Colliding hyperdimensional branes

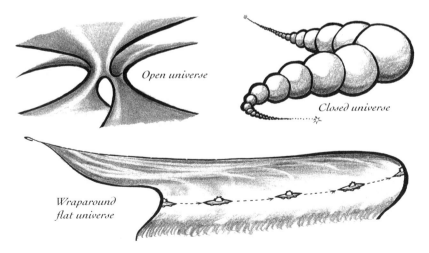

Open universe

Closed universe

Wraparound flat universe

COSMIC MYSTERIES
missing matter and other conundra

It is to be expected that a universe as diverse as ours may have the odd surprise hidden up its many-light-years-long sleeve. One question is why there is slightly more matter than antimatter, as theoretically both are created equally. Nature has a bias, which is difficult to explain (*top left*). Also we don't yet know why stuff has mass; is it due to an elusive Higgs boson, or is there a whiff of some new physics waiting beyond?

Observations of galactic motions hint that something mysterious is gluing everything together. To keep gravitational models on track, theorists invented *dark matter*, implying undiscovered exotic particles, massive neutrinos, and other gremlins. A dark matter weave has been proposed as seeding the large-scale bubbly cosmic fabric. Also puzzling is the fact that the redshift of distant supernovæ indicate the expansion of the universe may be accelerating. Space is being pushed apart by an enigmatic *dark energy*, possibly indicating a murky *cosmological constant* or the presence of eerie antigravity-causing antimatter in the great voids (*top right*).

Another head-scratcher involves the possibility that the physical constants might not be quite so constant. *Alpha* (*α*), the fine structure constant, which relates to electromagnetic interactions and ultimately whether living beings can exist, could have been different in other epochs. Surprisingly, it may also change across the universe (*lower left*).

Even this is small fry compared to *dark flow*, which seems to be pulling vast numbers of galaxy clusters in one direction, suggesting something incredibly massive outside the edge of our light bubble (*lower right*). If the evidence is to be believed, perhaps the universe behaves very differently indeed in distant parts of it we cannot even see.

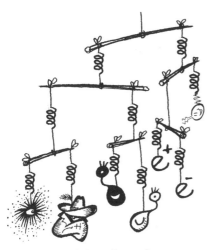

The universal balance tilts toward matter over antimatter, there is a chance that antimatter galaxies may exist, but none have as yet been found.

In the dark? The mass-energy of the universe seems to consist of dark energy 72%, dark matter 23%, ordinary atoms 4.6%, and less than 1% neutrinos.

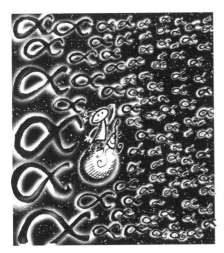

Capricious constants: Alpha, the fine structure constant, seems to vary depending on direction, perhaps as a result of a universal dipole.

Squillions of galaxies heading off at nearly two million mph toward Centaurus and Vela hint at a sibling universe beyond our event horizon.

HABITABLE ZONES
life, the universe, and everything

Life could be a rare and precious thing in the universe. For evolution to occur here on Earth we have needed a planet big enough to hold an atmosphere, close enough to the Sun for water to be liquid, and yet far enough away to avoid an inferno. The width of the comfortable circumstellar habitable zone has shrunk over time (*opposite top left*) as the Sun has become brighter, and in a billion years or so the Earth will be too hot for life as the Sun continues its expansion into a red giant.

Our mid-way position in the galactic disk is another crucial balancing act (*opposite top right*). Life needs time to evolve, plus the heavy elements synthesized in supernovæ. Although element-rich explosions happen regularly around the galactic center, they have the unfortunate side effect of sterilizing enormous areas of space. Conversely, the quiet outer edge of the disk lacks heavy elements, making life there equally unlikely.

Should a species survive long enough to form an intelligent society, there are several stages to look forward to. A *Type 0* society is divided by all-too familiar squabbles and inefficient use of resources, yet with international consensus, global communication, and renewable energy it can rise to a *Type I* status. Planetary *Type II* civilizations spread to neighboring worlds, while interstellar travel grants the next giant leap to galactic *Type III* civilizations. Eventually, with resources on a cosmic scale, *Type IV* beings can manipulate matter and spacetime.

Our solar system is not alone. Surveys suggest that there may be hundreds of billions across the galaxy, some with fast-orbiting planets with masses greater than Jupiter, others circling the two stars of a binary. Orphan planets wander in deep space. Life could be anywhere.

Circumstellar habitable zone

Galactic habitable zone

A FINELY TUNED UNIVERSE
strangely perfect for life

One of the most thought-provoking conundrums in modern cosmology goes by the name of the *Anthropic Principle*. This states that however you look at it, our universe appears to be *very* finely tuned to maximize the chances for biological life like us. Here are some examples:

If the force of gravity were 0.1.% stronger than it is now then the universe would just be full of black holes; if it were 0.1% weaker then no galaxies would have formed. If the Big Bang had exploded with a tiny amount less energy than it did, then the early universe would have collapsed in on itself; a little more energy and it would have expanded too fast for even stars to form. If the electromagnetic force was any weaker relative to the gravitational force then stars would collapse long before life had a chance to evolve. If the proton was slightly more massive than the neutron then hydrogen would decay, and most matter in the universe would decompose. If the strong nuclear force was any stronger then no atoms could have formed. Without its special energy level, insufficient carbon would be manufactured in stars for life to exist. If the properties of water were any different then ... the list goes on.

There are currently 25 fundamental dimensionless constants required for the standard model of physics, and the ratios between them govern the fundamental interactions of physics, and also the size, age, and expansion of the universe. So here's the big question—how did the universe spring into being so perfectly tuned for life like this?

Some scientists think this is a nova in a teacup. "Of course it's right for life, otherwise we wouldn't be here". But others are not so sure ...

Carbon-based life needs a specific fusion
reaction for carbon to occur in the stars.

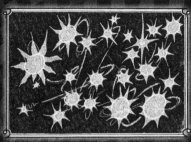
A weaker electromagnetic force would lead
to short-lived stars with no chance for life.

If gravity was stronger the universe would
just be full of supermassive black holes.

If there were equal quantities of matter
and antimatter then nothing would exist.

The Multiverse solution to the Anthropic Conundrum: There are squillions of
universes, each with their own mix of constants. We live in the one suited for us!

377

POSSIBLE EXPLANATIONS
for a finely tuned universe

The chances of the universe just "getting it right for life" like it did from the outset are so infinitesimally small that some scientists have started looking for an explanation of *how* it did it. Here are some ideas:

Like quantum mechanics and the popular Many Worlds theory (*see page 174*), various boffins have proposed that there must be billions of different random universes, and we happen to live in one where it all works beautifully for clever monkeys like us. Detractors from this (and the Many Worlds) claim that it defies Occam's razor (i.e. it invents colossal amounts of new and unprovable stuff), and so is bad science.

Others think there must be a *reason* for the fine tuning. Perhaps physics just pans out this way. More paranoid professors have suggested that the entire universe must be a virtual reality computer simulation of some kind. We're in a game, how else could it be so well set up? But then who wrote the program? This is actually the same as saying "God did it", or another Designer. Or maybe aliens did it ...

Neodarwinians, eager not to be left out of the fray, have advanced the idea of a biocosm. The reason the universe looks so perfect for life, they suggest, is because it is the child of successful parents. Highly evolved parent universes can pass on this perfect mix of constants, like DNA, to baby universes via a Big Bang.

Again, similar to quantum theories, there are solutions which suggest that some kind of panconscious reverse temporal causation is involved. Perhaps the universe really is a huge conscious mind, holographic and entangled, and like a quantum computer it just tried everything and came up with the best solution, right from the word go.

ULTIMATE REALITY
the kaleidoscopic playpen

With each passing second our view of the universe grows slightly wider, yet we still have little idea of how big it is, whether it has an edge, or whether ours is one of an infinitude, with dimensions unimagined.

Are we alone, and if not, who and what might our neighbours be? Or even what are we? Is a homely planet necessary, or does evolution operate in more extreme niches? On the very smallest scales today the distinction between matter and energy is blurred, leaving seemingly solid reality a phantasm. Curious subatomic entanglements separated by space and time suggest a holistic, unified universe, possibly itself the origin of biological consciousness. Maybe life is part of a greater series of interconnections, subtly guided by an underlying transtemporal panpsychic awareness. Stranger still, we may be just scintillating illusions, extrapolated from deeper realities (*e.g. opposite after M C Escher*).

Every page of this book, often every paragraph, and sometimes every sentence, can represent the work of many hardworking scientists, thinkers, and artists. Why the universe is even comprehensible is perhaps one of the greatest mysteries; after all, it need not necessarily have come into being behaving according to the simple laws and principles they have uncovered and which form the backbone of this book. Hopefully these pages will have amazed, informed, and inspired you, to look again at the world around you, and marvel.

Humanity has only just peeped out from beneath its earthly blindfold and taken the first stumbling steps off-world. As a species we are facing many challenges, not least the ones we have created, yet our very atoms were born in the stars, and one day we might return to visit.

APPENDICES
& INDEX

CONSTANTS

Speed of Light	*c*	670,616,629 mph, 186,282.4 miles s^{-1}, 2.99792458 x 10^8 m s^{-1}
Light-year	*ly*	0.306601 pc, 5,878,625,373,183.608 miles, 9,460,730,472,580 km
Parsec	*pc*	3.26156 ly, 1.9173508 x 10^{13} miles, 3.085678 x 10^{13} km
Astronomical Unit	*AU*	1.5813 x 10^{-5} ly, 92,955,791 miles, 149,597,892 km

Speed of Sound   *in dry air at 0°C* 331.4 m s^{-1},   1087.3 feet s^{-1} *in water at 20°C* 1482 m s^{-1},   4862 feet s^{-1}

Electron Mass	m_e	9.1093897 × 10^{-31} kg		*Gas constant*	*R*	8.314472 J K^{-1} mol^{-1}
		= 0.51099 MeV		*Gravitational const.*	*G*	6.674259 x 10^{-11} m^3 kg^{-1} s^{-2}
Electron charge	*e*	1.602176 × 10^{-19} C		*H$_2$0 triple point*	T_{tpw}	273.16 K
		= 4.803 × 10^{-10} esu		*Imped. of vacuum*	Z_0	376.7303 Ω
Proton mass	m_p	1.672623 × 10^{-27} kg		*Loschmidt const.*	n_0	2.6867775 x 10^{25} m^{-3}
		= 1836.1 x electron mass		*Mag. flux quantum*	ϕ_0	2.067833758 x 10^{-15} Wb
Neutron mass	m_n	1.6749286 x 10^{-27} kg		*Magnetic const.*	μ_0	4π x 10^{-7} NA^{-2}
Atomic mass unit	*u*	1.66054 x 10^{-27} kg		*Mol. vol. gas*	V_m	2.2413968 x 10^{-2} m^3 mol^{-1}
Atm. mass eng. eqv.	$m_u c^2$	931.494 MeV		*Perm. of vacuum*	μ_0	4π x 10^{-7} H m^{-1}
Avogadro's no.	N_A	6.02214 x 10^{23} mol^{-1}		*Permittivity const.*	ε_0	8.8541878 x 10^{-12} F m^{-1}
Bohr radius	a_0	5.2917725 x 10^{-11} mol^{-1}		*Planck constant*	*h*	6.62606896 x 10^{-34} J s
Boltzmann const.	*k*	1.38065 x 10^{-23} J K^{-1}		*Planck const/2π*	*ħ*	1.054571726 x 10^{-34} J s
Earth Escape Veloc.	v_{esc}	6.96 miles s^{-1}, 11.2 kms^{-1}		*Planck length*	l_P	1.616199 x 10^{-35} m
Earth Grav. Accel.	g_n	32.174 feet s^{-2},		*Planck time*	t_P	5.31906 x 10^{-44} s
		9.80665 ms^{-2}		*Planck temp*	T_P	1.41683 x 10^{32} K
Earth Std. Atmos.		101,325 Pa		*Planck mass*	m_P	2.17651 x 10^{-8} kg
Faraday constant	*F*	9.648534 x 10^4 C mol^{-1}		*Rcp. fine str. const.*	$1/\alpha$	137.035999
Fine struct. const.	*α*	7.297353 x 10^{-3}		*Rydberg const.*	R_H	1.097373 x 10^7 m^{-1}

Hydrogen Emission Spectra:

Lyman series (ultraviolet)	*Balmer series (visible)*	*Paschen series (infrared)*	*Brackett series (infrared)*
	n = 2 λ = 3650 Å	n = 3 λ = 8210 Å	n = 4 λ = 14,592 Å
n = 1 λ = 912 Å	n = 3 λ = 6560 Å	n = 4 λ = 18,761 Å	n = 5 λ = 40,532 Å
n = 2 λ = 1216 Å	n = 4 λ = 4860 Å	n = 5 λ = 12,830 Å	n = 6 λ = 26,300 Å
n = 3 λ = 1026 Å	n = 5 λ = 4340 Å		
n = 4 λ = 973 Å	n = 6 λ = 4100 Å		

log$_{10}$ metres	-20	-15	-10	-5	0	5	10	15	20	25

Quarks Nucleus   Molecule   Sand Mouse Whale Asteroid Star   Solar System   Galaxy   Galactic Wall

Proton   Atom   Cell   Insect Human Forest Planet   Super Giant   Glob.Cluster   Super Cluster

UNITS

Physical Quantity	Symbol for quantity	Name of SI unit	Unit symbol	Immediate definition	Basic units definition	As mass, length, time, & current
frequency	f	hertz	Hz	s^{-1}	s^{-1}	T^{-1}
force	F	newton	N	$kg\ m\ s^{-2}$	$kg\ m\ s^{-2}$	$M\ L\ T^{-2}$
energy	W	joule	J	$N\ m$	$kg\ m^2\ s^{-2}$	$M\ L^2\ T^{-2}$
power	P	watt	W	$J\ s^{-1}$	$kg\ m^2\ s^{-3}$	$M\ L^2\ T^{-3}$
pressure	p	pascal	Pa	$N\ m^{-2}$	$kg\ m^{-1}\ s^{-2}$	$M\ L^{-1}\ T^{-2}$
charge	Q	coulomb	C	$A\ s$	$A\ s$	$T\ I$
voltage	V	volt	V	$J\ C^{-1}$	$kg\ m^2\ s^{-3}\ A^{-1}$	$M\ L^2\ T^{-3}\ I^{-1}$
capacitance	C	faraday	F	$C\ V^{-1}$	$A^2\ s^4\ kg^{-1}\ m^{-2}$	$M^{-1}\ L^{-2}\ T^4\ I^2$
resistance	R	ohm	Ω	$V\ A^{-1}$	$kg\ m^2\ s^{-3}\ A^{-2}$	$M\ L^2\ T^{-3}\ I^{-2}$
conductance	G	siemens	S	Ω^{-1}	$kg^{-1}\ m^{-2}\ s^3\ A^2$	$M^{-1}\ L^{-2}\ T^3\ I^2$
flux density	B	tesla	T	$N\ A^{-1}\ m^{-1}$	$kg\ s^{-2}\ A^{-1}$	$M\ T^{-2}\ I^{-1}$
magnetic flux	Φ	weber	Wb	$T\ m^2$	$kg\ m^2\ s^{-2}\ A^{-1}$	$M\ L^2\ T^{-2}\ I^{-1}$
inductance	L	henry	H	$V\ A^{-1}\ s$	$kg\ m^2\ s^{-2}\ A^{-2}$	$M\ L^2\ T^{-2}\ I^{-2}$

HADRONS

Baryons	Symbol	Mass	Quarks	Charge	Spin
Proton	N^+	938	uud	+1	$1/2$
Neutron	N°	940	ddu	0	$1/2$
Sigma$^+$	Σ^+	1198	uus	+1	$1/2$
Sigma$^\circ$	Σ°	1192	dus	0	$1/2$
Sigma−	S^-	1197	dds	−1	$1/2$
Lambda$^\circ$	Λ°	1116	dus	0	$1/2$
Xi$^\circ$	Ξ°	1315	uss	0	$1/2$
Xi$^-$	Ξ^-	1321	dss	−1	$1/2$
Sigma$^+$	Σ^+	938	uus	+1	$1/2$
Delta^{++}	Δ^{++}	1231	uuu	+2	$3/2$
Delta$^+$	Δ^+	1232	duu	+1	$3/2$
Delta$^\circ$	Δ°	1234	ddu	0	$3/2$
Delta$^-$	Δ^-	1235	ddd	−1	$3/2$
Sigma^{*+}	Σ^{*+}	1189	uus	+1	$3/2$
Sigma$^{*\circ}$	$\Sigma^{*\circ}$	1193	dus	0	$3/2$
Sigma^{*-}	Σ^{*-}	1197	dds	−1	$3/2$
Xi$^{*\circ}$	$\Xi^{*\circ}$	1315	uss	0	$3/2$
Xi^{*-}	Ξ^{*-}	1321	dss	−1	$3/2$
Omega$^-$	Ω^-	1672	sss	−1	$3/2$

Mesons	Symbol	Mass	Quarks	Charge	Spin
Pi$^+$	π^+	140	$u\bar{d}$	+1	0
Pi$^\circ$	π°	135	$u\bar{u},\ d\bar{d}$	0	0
Pi$^-$	π^+	140	$d\bar{u}$	−1	0
Eta	η^+	547	$u\bar{u},\ d\bar{d},\ s\bar{s}$	0	0
Eta prime	η'	958	$u\bar{u},\ d\bar{d},\ s\bar{s}$	0	0
Kaon$^+$	K^+	494	$u\bar{s}$	+1	0
Kaon$^\circ$	K°	498	$d\bar{s}$	0	0
Rho$^+$	ρ^+	770	$u\bar{d}$	+1	1
Rho$^\circ$	ρ°	770	$u\bar{u},\ d\bar{d}$	0	1
Omega	ω	782	$u\bar{u},\ d\bar{d}$	0	1
Phi	ϕ	1020	$s\bar{s}$	0	1
K^{*+}	K^{*+}	892	$u\bar{s}$	+1	1
K$^{*\circ}$	$K^{*\circ}$	892	$d\bar{s}$	0	1
J/ψ	ψ	3097	$c\bar{c}$	0	1
Upsilon	Υ	9460	$b\bar{b}$	0	1

Note that over 200 baryons and 30 mesons have been discovered.

WEIGHTS & MEASURES

British Standard Imperial Weights & Measures

1 yard = 3 feet = 36 inches
1 furlong = 220 yards = 660 feet
1 nautical mile = 6080 feet
1 pound = 16 ounces
1 hundredweight = 112 pounds
1 pint = 4 gills = 20 fluid ounces

1 fathom = 2 yards = 6 feet
1 mile = 1760 yards = 5280 feet
1 acre = 4840 square yards
1 stone = 14 pounds
1 ton = 2240 pounds
1 gallon = 277.4 cubic inches

British Imperial : Metric

1 inch = 2.54000 cm
1 foot = 0.304800 meter
1 mile = 1.60934 kilometers
1 ounce = 28.3495 grammes
1 pound = 0.45359237 kilogram
1 (long) ton = 1016.047 kg
1 gallon = 4.549631 litres
1 bushel = 36.397 litres
1 acre = 0.404687 hectares
1 cubic inch = 16.3871 cubic cm
°fahrenheit = $\frac{9}{5}$(°celsius) + 32
1 horsepower = 0.7457 kilowatts
1 pound per sq. inch = 0.0688 atm
1 foot pound = 1.355 joules

Metric : British Imperial

1 cm = 0.393701 inches
1 meter = 3.280842 feet
1 kilometer = 0.621371 miles
1 gramme = 0.0352740 ounces
1 kilogram = 2.204622 pounds
1 tonne = 0.9842064 (long) tons
1 litre = 0.219975 gallons
1 litre = 0.027475 bushels
1 hectare = 2.47105 acres
1 cubic cm = 0.0610237 cubic inches
°celsius = $\frac{5}{9}$(°fahrenheit − 32)
1 kilowatt = 1.3410 horsepower
1 atmosphere = 14.696 lbs/sq. inch
1 joule = 0.738 foot pounds

US Weights & Measures differing from UK

1 US hundredweight = 100 pounds
1 US (short) ton = 0.892857 UK (long) tons
1 US gallon = 3.785412 litres
1 US gallon = 0.83267 UK gallons
1 dry gallon = 4.404884 litres
1 US bushel = 35.2391 litres
1 US bushel = 0.9689 UK bushels

1 US (short) ton = 2000 pounds
1 US (short) ton = 907.184 kilogram
1 litre = 0.264172 US gallons
1 UK gallon = 1.20095 US gallons
1 litre = 0.227021 dry gallons
1 litre = 0.028378 US bushels
1 UK bushel = 1.0321 U.S. bushels

General Metric

°kelvin (K) = °celsius + 273.15
1 atmosphere = 101325 Pa
1 calorie (cal) = 4.184 J
1 esu = 3.3356 × 10^{-10} C (coulombs)
1 erg = 2.390 × 10^{-11} kcal
1 eV / molecule = 96.485 kJ mol^{-1}
1 kcal mol^{-1} = 349.76 cm^{-1}, 0.0433 eV
1 newton per square meter (N m^{-2}) = 1 Pascal (Pa)

1 angström (A) = 10 × 10^{-10} m
1 bar = 105 Pa
1 curie (Ci) = 3.7 × 10^{10} s^{-1}
1 erg = 10^{-7} J
1 eV = 1.60218 × 10^{-19} J
1 joule (J) = 0.2389 cal
1 kJ mol^{-1} = 83.54 cm^{-1}
1 wave no. (cm^{-1}) = 2.8591 × 10^{-3} kcal mol^{-1}

Prefixes: 10^3 = *kilo* 10^6 = *mega* 10^9 = *giga* 10^{12} = *tera* 10^{15} = *peta* 10^{18} = *exa*
10^{-3} = *milli* 10^{-6} = *micro* 10^{-9} = *nano* 10^{-12} = *pico* 10^{-15} = *femto* 10^{-18} = *atto*

Expansions & Extras

$\pi = 3.14159265358979\ldots$ $\qquad$ $e = 2.718281828459045\ldots$

$\sqrt{2} = 1.414213562373095\ldots$ $\qquad$ $\sqrt{\pi} = 1.7724538500905516\ldots$

$$e^x = 1 + x + \frac{x^2}{2!} + \frac{x^3}{3!} + \frac{x^4}{4!} + \ldots, \quad \text{so} \quad e = 1 + 1 + \frac{1}{2!} + \frac{1}{3!} + \frac{1}{4!} + \ldots$$

$$\log(1+x) = x - \frac{x^2}{2} + \frac{x^3}{3} - \frac{x^4}{4} + \ldots \;(-1 < x < 1)$$
$$e = 2 + \cfrac{1}{1+\cfrac{1}{2+\cfrac{1}{1+\cfrac{1}{1+\cfrac{1}{4+\cfrac{1}{1+\ddots}}}}}}$$

$$\sqrt{2} = 1 + \cfrac{1}{2+\cfrac{1}{2+\cfrac{1}{2+\cfrac{1}{2+\cfrac{1}{2+\ddots}}}}}$$
$$\sqrt{3} = 1 + \cfrac{1}{1+\cfrac{1}{2+\cfrac{1}{1+\cfrac{1}{2+\cfrac{1}{1+\ddots}}}}}$$
$$\phi = 1 + \cfrac{1}{1+\cfrac{1}{1+\cfrac{1}{1+\cfrac{1}{1+\ddots}}}}$$

$$\frac{1}{1-x} = 1 + x + x^2 + x^3 + x^4 + \ldots \;(-1 < x < 1)$$

$$\pi = 4\left(\frac{1}{1} - \frac{1}{3} + \frac{1}{5} - \frac{1}{7} + \frac{1}{9} - \ldots\right)$$
$$\pi = 3 + \cfrac{1^2}{6+\cfrac{3^2}{6+\cfrac{5^2}{6+\cfrac{7^2}{6+\cfrac{9^2}{6+\cfrac{11^2}{6+\cfrac{13^2}{6+\ddots}}}}}}}$$

$$\text{arcsin } x = x + \frac{1}{2}\frac{x^3}{3} + \frac{1}{2}\frac{3}{4}\frac{x^5}{5} + \frac{1}{2}\frac{3}{4}\frac{5}{6}\frac{x^7}{7} + \ldots \quad radians$$

$$\sin x = x - \frac{x^3}{3!} + \frac{x^5}{5!} - \frac{x^7}{7!} + \ldots, \quad \cos x = 1 - \frac{x^2}{2!} + \frac{x^4}{4!} - \frac{x^6}{6!} + \ldots \quad (x \text{ in radians})$$

Taylor expansion: $\; f(x) = f(x-a) + af'(x-a) + \frac{a^2}{2!}f''(x-a) + \frac{a^3}{3!}f'''(x-a) + \ldots$

Maclaurin expansion: $\; f(x) = f(0) + xf'(0) + \frac{x^2}{2!}f''(0) + \frac{x^3}{3!}f'''(0) + \ldots$

Fractions: $\quad \dfrac{a}{b} + \dfrac{c}{d} = \dfrac{ad+bc}{bd} \qquad \dfrac{a}{b} - \dfrac{c}{d} = \dfrac{ad-bc}{bd} \qquad \dfrac{a}{b} \times \dfrac{c}{d} = \dfrac{ac}{bd} \qquad \dfrac{a}{b} \div \dfrac{c}{d} = \dfrac{ad}{bc}$

'Tan $\frac{1}{2}$' formulæ: if $t = \tan\frac{1}{2}\theta$, $\quad \sin\theta = \dfrac{2t}{1+t^2} \quad \cos\theta = \dfrac{1-t^2}{1+t^2} \quad \tan\theta = \dfrac{2t}{1-t^2}$

Riemann's zeta function: "*Probably the most challenging and mysterious object of modern mathematics, in spite of its utter simplicity*" M.C. Gutzwiller

$$\zeta(x) = 1 + \frac{1}{2^x} + \frac{1}{3^x} + \frac{1}{4^x} \ldots = \left(\frac{1}{1-\frac{1}{2^x}}\right)\left(\frac{1}{1-\frac{1}{3^x}}\right)\left(\frac{1}{1-\frac{1}{5^x}}\right)\cdot \quad \left(\frac{1}{1-\frac{1}{p_k^x}}\right) \quad (x > 1)$$

where p_k is the kth prime number.

CARBON CHEMISTRY

Group Name	Functional Group	Suffix Name	First Member Formula	Name	General Formula
Alkane	$R-CH_3$	-ane	$H-\overset{H}{\underset{H}{C}}-H$	methane	C_nH_{2n+2}
Alkene	$\overset{H}{\underset{R'}{\,}}C=C\overset{H}{\underset{R}{\,}}$	-ene	$\overset{H}{\underset{H}{\,}}C=C\overset{H}{\underset{H}{\,}}$	ethene	C_nH_{2n}
Alkyne	$R-C\equiv C-R'$	-yne	$H-C\equiv C-H$	ethyne	C_nH_{2n-2}
Alcohol	$R-\overset{OH}{\underset{H}{C}}-R'$	-anol	$H-\overset{OH}{\underset{H}{C}}-H$	methanol	$C_nH_{2n+1}OH$
Aldehyde	$R-\overset{O}{C}-H$	-anal	$H-\overset{O}{C}-H$	methanal	$C_nH_{2n-1}OH$
Ketone	$R\underset{O}{\diagdown}\overset{\diagup CH_3}{\,}$	-anone	$H\underset{O}{\diagdown}\overset{\diagup CH_3}{\,}$	methanone	$C_nH_{2n}O$
Carboxylic acid	$R-\underset{O}{C}-OH$	-anoic acid	$H-\underset{O}{C}-OH$	methanoic (formic) acid	
Amine	H_2N-R	-anamine	$CH_3-N\overset{H}{\underset{H}{\diagdown}}$	methylamine	
Ester	$R\overset{O}{\diagup}C\diagdown_{OR'}$	-ate	$CH_3-C\overset{\diagup O}{\diagdown OCH_3}$	carboxylic acid + alcohol	
Amide	$R\overset{O}{\diagup}C\diagdown_{NHR'}$	-ide	$CH_3\overset{O}{C}\diagdown_{NHR'}$	carboxylic acid + amine	
Ether	$R-O-R'$	-oxy... -ane	CH_3O-CH_3	methoxy methane	
Cyclopropane		Benzene		or ⬡ or ⬡	

Branched alkanes:
branch name changes from -ane to -yl.

ethyl branch — 3-ethyl hexane

methyl branch — 2-methyl pentane

Prefix (no. of carbon atoms): Meth-1, Eth-2, Prop-3, But-4, Pent-5, Hex-6, Hept-7, Oct-8, Non-9, Dec-10
The symbols R and R' denote either a hydrogen atom, or a hydrocarbon side chain.

EXAMPLES OF ELECTRON ORBITALS

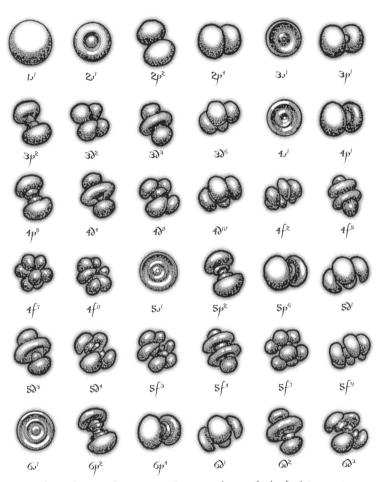

$1s^1$	$2s^1$	$2p^2$	$2p^4$	$3s^1$	$3p^1$
$3p^2$	$3d^2$	$3d^3$	$3d^6$	$4s^1$	$4p^1$
$4p^5$	$4d^4$	$4d^8$	$4d^{10}$	$4f^2$	$4f^5$
$4f^7$	$4f^{11}$	$5s^1$	$5p^2$	$5p^6$	$5d^1$
$5d^3$	$5d^4$	$5f^3$	$5f^4$	$5f^7$	$5f^9$
$6s^1$	$6p^2$	$6p^4$	$6d^1$	$6d^2$	$6d^3$

These enclosure surfaces represent the various electron orbitals of a hydrogen atom.
Each traces the quantum probability wavefunction of an electron pair as they juggle with the many
attractive and repulsive forces inside the atom to give a fantastically labyrinthine interweave.
S orbitals are spherical. p orbitals have twin lobes. d orbitals are four-fold or enjoy
a single donut, whilst double donuts and six lobed shapes belong to the f orbitals.

The Periodic Table

Group	IA (1)	IIA (2)		IIIB (3)	IVB (4)	VB (5)	VIB (6)	VIIB (7)	VIII (8)
Period 1	H 0 — 1 **H** Hydrogen 1.00079 1310								
2	B 4 — 3 **Li** Lithium 6.941 519	H 5 — 4 **Be** Berylium 9.01218 900							
3	B 12 — 11 **Na** Sodium 22.9878 494	H 12 — 12 **Mg** Magnesium 24.3050 736							
4	B 20 — 19 **K** Potassium 39.0983 418	C 20 — 20 **Ca** Calcium 40.0785 590		H 24 — 21 **Sc** Scandium 44.9559 632	H 26 — 22 **Ti** Titanium 47.867 661	B 28 — 23 **V** Vanadium 50.9415 648	B 28 — 24 **Cr** Chromium 51.9961 653	C 30 — 25 **Mn** Manganese 54.9380 716	B 30 — 26 **Fe** Iron 55.845 762
5	B 48 — 37 **Rb** Rubidium 85.4678 402	F 50 — 38 **Sr** Strontium 87.62 548		H 50 — 39 **Y** Yttrium 88.9059 636	B 50 — 40 **Zr** Zirconium 91.224 669	B 52 — 41 **Nb** Niobium 92.9064 653	B 56 — 42 **Mo** Molybdenum 95.94 694	H 55 — 43 **Tc** Technetium 97.9072 699	H 58 — 44 **Ru** Ruthenium 101.07 724
6	B 78 — 55 **Cs** Caesium 132.905 376	B 82 — 56 **Ba** Barium 137.327 502	57 - 70 Lanthanide series *	H 104 — 71 **Lu** Lutetium 174.967 481	H 108 — 72 **Hf** Hafnium 178.49 531	B 108 — 73 **Ta** Tantalum 180.948 760	B 110 — 74 **W** Tungsten 183.84 770	H 112 — 75 **Re** Rhenium 186.207 762	H 116 — 76 **Os** Osmium 190.23 841
7	H 136 — 87 **Fr** Francium 223.02 381	H 138 — 88 **Ra** Radium 226.025 510	89 - 102 Actinide series **	? 157 — 103 **Lr** Lawrencium 262.110 444	? 153 — 104 **Rf** Rutherfordium 263.113 490	? 157 — 105 **Db** Dubnium 262.114 ?	? 157 — 106 **Sg** Seaborgium 266.122 ?	? 157 — 107 **Bh** Bohrium 264.125 740	? 161 — 108 **Hs** Hassium 269.134 ?

* Lanthanide series

H 82 — 57 **La** Lanthanum 138.906 540	C 82 — 58 **Ce** Cerium 140.116 665	H 82 — 59 **Pr** Praseodymium 140.908 556	H 82 — 60 **Nd** Neodymium 144.24 607	H 84 — 61 **Pm** Promethium 144.913 556	R 90 — 62 **Sm** Samarium 150.36 540	B 90 — 63 **Eu** Europium 151.964 548	H 94 — 64 **Gd** Gadolinium 157.25 594	H 94 — 65 **Tb** Terbium 158.925 648

** Actinide series

C 138 — 89 **Ac** Actinium 227.028 669	C 142 — 90 **Th** Thorium 232.038 674	T 140 — 91 **Pa** Proctactinium 231.0356 568	O 146 — 92 **U** Uranium 238.029 385	O 144 — 93 **Np** Neptunium 237.048 604	M 150 — 94 **Pu** Plutonium 244.064 585	H 148 — 95 **Am** Americium 243.061 578	H 151 — 96 **Cm** Curium 247.070 581	H 150 — 97 **Bk** Berkelium 247.070 601

IUPAC interim naming system for new elements : o-nil-(n), 1-un-(u), 2-bi-(b), 3-tri-(t), 4-quad-(q), 5-pent-(p), 6-hex-(h), 7-sept-(s), 8-oct-(o), 9-enn-(e)

OF THE ELEMENTS

									H 2 **2** **He** Helium 4.0026 2370

				R 6 **5** **B** Boron 10.811 799	H 6 **6** **C** Carbon 12.0107 1090	H 7 **7** **N** Nitrogen 14.0067 1400	M 8 **8** **O** Oxygen 15.9994 1310	M 10 **9** **F** Fluorine 18.9984 1680	C 10 **10** **Ne** Neon 20.1797 2080
				C 14 **13** **Al** Aluminium 26.9815 577	C 14 **14** **Si** Silicon 28.0855 786	TC 16 **15** **P** Phosphorus 30.9738 1060	O 16 **16** **S** Sulphur 32.066 1000	O 18 **17** **Cl** Chlorine 35.4527 1260	C 22 **18** **Ar** Argon 39.948 1520
H 32 **27** **Co** Cobalt 58.9332 757	C 30 **28** **Ni** Nickel 58.6934 736	C 34 **29** **Cu** Copper 63.546 745	H 34 **30** **Zn** Zinc 65.39 908	O 38 **31** **Ga** Gallium 69.723 577	C 42 **32** **Ge** Germanium 72.61 762	R 42 **33** **As** Arsenic 74.9216 966	M 46 **34** **Se** Selenium 78.96 941	O 44 **35** **Br** Bromine 79.904 1140	C 48 **36** **Kr** Krypton 83.80 1350
C 58 **45** **Rh** Rhodium 102.906 745	C 60 **46** **Pd** Palladium 106.42 803	C 60 **47** **Ag** Silver 107.868 732	H 66 **48** **Cd** Cadmium 112.411 866	T 66 **49** **In** Indium 114.818 556	T 70 **50** **Sn** Tin 117.710 707	R 70 **51** **Sb** Antimony 121.760 833	R 78 **52** **Te** Tellurium 127.60 870	O 74 **53** **I** Iodine 126.904 1010	C 78 **54** **Xe** Xenon 131.29 1170
C 116 **77** **Ir** Iridium 192.217 887	C 117 **78** **Pt** Platinum 195.078 866	C 118 **79** **Au** Gold 196.967 891	R 122 **80** **Hg** Mercury 200.59 1010	H 124 **81** **Tl** Thallium 204.383 590	C 126 **82** **Pb** Lead 207.2 716	M 126 **83** **Bi** Bismuth 208.980 703	C 126 **84** **Po** Polonium 208.982 812	? 125 **85** **At** Astatine 209.987 920	C 136 **86** **Rn** Radon 222.017 1040
? 159 **109** **Mt** Meitnerium 268.139 ?	? 162 **110** **Ds** Darmstadtium 272.146 ?	? 161 **111** **Rg** Roentgenium 272.154 ?	? 165 **112** **Cn** Copernicium 277 ?	? 173 **113** **Uut** Ununtrium 286 ?	? 175 **114** **Uuq** Ununquadium 289 ?	? 174 **115** **Uup** Ununpentium 289 ?	? 177 **116** **Uuh** Ununhexium 293 ?	? 177 **117** **Uus** Ununseptium 294 ?	? 176 **118** **Uuo** Ununoctium 294 ?

H 98 **66** **Dy** Dysprosium 162.50 657	H 98 **67** **Ho** Holmium 164.930 ?	H 98 **68** **Er** Erbium 167.26 ?	H 100 **69** **Tm** Thulium 168.934 ?	C 104 **70** **Yb** Ytterbium 173.04 598
HC 153 **98** **Cf** Californium 251.080 608	C 158 **99** **Es** Einsteinium 252.083 619	? 157 **100** **Fm** Fermium 257.095 627	? 157 **101** **Md** Mendelevium 258.098 635	? 157 **102** **No** Nobelium 259.101 642

Crystal Structure   Number of neutrons
(See below for key)   (most abundant or stable isotope)

ATOMIC NUMBER

Chemical Symbol

Name of Element

Atomic Weight   First Ionization Energy
(Average relative mass)   (kJ mol⁻¹)

B body centred cubic   C cubic close packing   H hexagonal close packing   M monoclinic   O orthorhombic   R rhombohedral (trigonal)   T tetragonal   TC triclinic

WATER, MINERALS, & FLUID COMPARTMENTS

When biologists are asked what sort of life we have on Earth, the reply used to be 'carbon based', but now it is 'water based'. Water's unique properties allow all the biochemical and electrical transformations in life to take place, and it is the main constituent of your body. Body fluids come in three major compartments: blood, *extracellular fluid* (ECF, the salty soup between cells), and *intracellular fluid* (ICF, the viscous 'cytosol' inside cells). Nearly half of your blood volume is cells, mainly red cells, and a percent or two white cells and platelets.

The fluid portion of blood, *plasma*, is quite similar to extracellular fluid, but very large proteins like albumin stay in the blood, contributing to its osmotic ability to keep water within capillaries. About 30 litres of fluid seep out of capillaries per day, and all but 3 litres (drained as lymph) seeps back in, carrying cell products.

Fluid ratios vary with body type, gender, and age, but an adult has about (in litres): BLOOD 5, ECF 9, ICF 29.

Most blood is 'pooled' as a slow flowing reservoir in veins, between carrying everything else around. Brain and kidneys take a fifth of (non-lung) flow each, and the busiest current organs take most of the rest, especially muscles (if you're running), gut and liver (after a meal), or skin, (if you're hot).

In terms of dissolved molecules (rather than ions), ICF composition varies hugely, depending on cell type and phase of activities. Plasma carries all circulating substances, and ECF mediates between ICF & plasma.

Charged ions are kept at stable levels in fluid compartments, because life processes operate within narrow local electrical and acid-base windows. An ion is surrounded by a multi-layered morphing 'aquahedron' (*i.*) of water molecules, defining its sphere of influence.

Potassium and sodium ions both carry a single positive charge, but though K$^+$ is heavier, it is less electro-dense, so has a smaller, less influential aquahedron. This makes it the tame cation of choice for intracellular use, and for stabilizing the membrane voltage by seeping through small cation channels. Sodium performs the faster, more electro-dense processes like rushing

through voltage-gated channels, and is ejected from cells by Na-K-ATP pumps, 3 Na$^+$ out to 2 K$^+$ in per ATP molecule. The salty sea of ECF has chloride (Cl$^-$) as its main anion, while phosphates and proteins carry most ICF negative charges.

MAJOR ION CONCENTRATIONS (mmolL-1)

	Plasma	ECF	ICF
Sodium	144	146	11
Potassium	4	4	148
Calcium	4	3	0.1–3
Magnesium	2	2	35
Chloride	104	118	4
Bicarbonate	25	27	13
Phosphate	2	2	40–100
Protein anions	20	1	52
Others	5	6	60

Of other minerals, iron (mainly for hæmoglobin) is more abundant in men, whereas copper is higher in women. Rarer elements (Cu, Mg, Mn, Zn, Cr, Se, etc.) are mainly enzyme components, for instance molybdenum, the rarest element you need, which is in xanthine oxidase. A cobalt sits in the middle of vitamin B12 (*ii.*), a co-enzyme, like most vitamins.

Sulfur appears mainly in proteins, playing a special role in forming their three-dimensional shape. Iodine is used in thyroxine synthesis, being super-added to iodine-containing peptides.

A fourth fluid category is special extracellular secretions, which includes the cerebro-spinal fluid the brain floats in, the eyes' aqueous and vitreous humours, and the ears' endolymph and perilymph, and also any fluid or mucus matrix cells deposit around them, for lubrication, insulation, protection, and as an extended molecular territory for cellular activities.

About nine-tenths of your water comes in what you eat and drink, but, incredibly, the other tenth is made by you as fresh new water molecules, a little during protein synthesis, but most of it during mitochondrial ATP synthesis.

MOLECULAR NUTRITION

WHAT YOU HAVE TO EAT & DRINK is: water and minerals, lipids (fats), carbohydrates (sugars), proteins, and vitamins. The complex molecules in your food have already been synthesized by the combined efforts of bacteria, fungi, plants, and animals (if you eat meat). The energy source for the synthesis of complex organic molecules is the Sun.

Your body can interconvert most substances, but some molecules the body cannot synthesize are 'essential' nutrients, mainly certain amino-acids, lipids and vitamins. Everything else can be made from other molecules.

Catabolism is the breakdown of complex molecules into simple ones that are used to supply energy or to build other compounds. *Anabolism* is the synthesis of complex compounds. These processes require energy input, which in most non-mitochondrial pathways comes as *ATP* (*a.*).

a. adenosine-triphosphate

The *Krebs cycle* is the multi-step circular pathway by which mitochondria produce *ATP*. The brain has to use *glucose* (*b.*) to fuel this process, and other cells prefer to, as it can be split via a simple eleven-step process into two *pyruvates* (e.g. *c.*) which enter the Krebs cycle two steps later. Most of the pathways of sugar, fat, and protein catabolism converge around here. Short-term lack of oxygen diverts pyruvate to *lactate* (*d.*) in an 'emergency' pathway.

Dietary sugars can be simple like glucose, or the fruit sugar *fructose* (*e.*), disaccharides like *sucrose* (*b.* + *e.*) or *lactose* (*f.*), the milk sugar which is the main non-plant carbohydrate you're likely to ingest, or complex *sugar polymers* (starches). Bean sugars are initially digested by gas-producing bacteria, hence the flatulence. Sugars can be stored as *glycogen*, a glucose polymer, synthesized by liver

and skeletal muscle cells as a rapid access reserve, but 99% of energy storage is as fat, mainly *triglycerides* e.g. (*g.*), the

g.

destination of all excess sugars, proteins, and lipids. The two essential dietary fats are *linoleic* and *linolenic acid* (*b.*) found

b.

in all plant cells. Other dietary fats are *cholesterol* (essential in cell walls and for making steroid hormones, bile salts & vitamin D), and *phospholipids* (like lecithin), the main constituents of cell walls. Derivatives of the 20-carbon arachidonic acid, made from linolenic acid, are also used as *paracrine* messengers between cells, both within tissues, and in moveable operations like damage repair and *immunication*.

j.   *R = radical*

Plants make all amino-acids (*j.*), and we can synthesize 10 of the basic 20, the rest being 'essential' amino-acids like *tryptophan* (*k.*). 'Peptide' bonds join amino-acids to make proteins of all sizes, from *dipeptides* (two amino-acids) to huge muscle proteins like *titin*, a chain of 27,000. If you have to use proteins for energy, you remove their nitrogen containing parts, creating

k.

ammonia, which is also made at many other metabolic stages. Birds excrete excess nitrogen as urates in semi-solid guano, but mammals excrete ammonia dissolved as *urea* (*l.*), which is the

l.

main reason you have to drink so much water. By the time you eat your food, the hard work of synthesis has already been done. Our metabolic processes have evolved with the organisms we eat, and use the same substances in different pathways, so a diet of good water, fruit, nuts, roots, vegetables and some extra protein, especially in childhood, will supply all your needs.

NEUROTRANSMITTERS

Nerves send their messages to other nerves or muscles using neurotransmitters (NTs). NTs tend to be small molecules, mainly amino-acid derivatives, which travel across the gaps in synapses or neuro-muscular junctions, binding to receptors on target cells, which opens ion channels (Na^+ and/or Ca^{2+} in excitatory neurons, and K^+ or Cl^- in inhibitory ones). Receptors come in several subtypes for most NTs, having variable or polar effects in different cells, and are often named after natural drugs that excite or block them, like nicotinic and muscarinic receptors for **ACETYL-CHOLINE** (*ACh*). *ACh* is the most 'pure' NT; all the others have some hormone-like phasic activity on certain cells, like **NORADRENALINE** (NA). ACh and NA are the main NTs of the peripheral nervous system, and can be excitatory or inhibitory, receptor-dependant. ACh and NA are also Central Nervous System NTs. Other major CNS NTs are:

SEROTONIN (5-HT); tryptophan derivative, gut neuro-hormone and rare but crucial midbrain/brainstem NT. Mood, humour, 'significance of perceptions' on all levels.

DOPAMINE; unlike 5-HT, mainly excitatory, midbrain motor, and limbic motivation/reward/satisfaction areas.

HISTAMINE; inhibitory in hypothalamic temperature & arousal circuits. Common peripheral inflammatory signaller.

GABA; prevalent local inhibitor in brain. Relaxation in general, as after a meal. Many addictive drugs mimic it.

GLYCINE; spinal cord 'GABA', blocked by strychnine.

GLUTAMATE & **ASPARTATE**; excite non-local integration.

ENDORPHINS & other **PEPTIDE OPIOIDS**; inhibitory, widespread, esp. pain tracts. Transitory blissful peace.

SUBSTANCE P; excites, 'anti-endorphin', chili, P for pain.

NITRIC OXIDE; bloodflow, and non-CNS local control of heart/artery/gut muscles. In nerves to penis and clitoris.

MAJOR HORMONES BY GLAND

'Hormone' is a loose term these days, with new messaging systems being discovered regularly. 'Proper' hormones are either amino-acid derivatives (as amines, polypeptides, or glycoproteins) or cholesterol derivatives (called 'steroids', made by the adrenal cortex, testis/ovary, and placenta).

PINEAL; releases many newly discovered peptides, but its only even partially understood product is *melatonin*, made from *serotonin* during darkness. Dual rhythms (*ps. 308 & 312*).

HYPOTHALAMUS; secretes only 'release' and 'inhibitory' *factors* controlling pituitary, via dedicated capillary loops.

ANTERIOR PITUITARY; controls lower glands with 'trophic' hormones, like *ACTH* (adrenal cortex), *TSH* (thyroid), and *FSH/LH* (ovaries & testes). Controls pigmentation (*MSH*), growth (*GH, p.312*), fatness (*LPH*), and milk production (*prolactin, p.310*).

POSTERIOR PITUITARY; makes two nonapeptide hormones (nine-amino-acids) which differ by just two amino-acids; *oxytocin*; 'letting go' of feelings, fetus, milk, and memories, and *vasopressin*; 'holding on' to water, memories & emotions.

THYROID; speed of development, metabolism, growth, and energy. *Thyroxines* are iodine-added peptides which travel into cells, and either bind with cell's *zinc-finger proteins* (off to the nucleus to regulate transcription rates of sets of 'master' genes), or short circuit mitochondrial ATP production, resulting in release of energy as heat.

PARATHYROID; *PTH* opposes thyroid's *calcitonin* (*p.310*)

THYMUS; childhood immune development (*p.298*)

PANCREAS; islet cells secrete a) *glucagon* (29 amino-acid peptide), catabolic king of gut hormones, which opposes b) *insulin* (two-unit peptide, 21+30 amino-acids) which stimulates storage, repair, & anabolism in general. Other pancreatic hormones control digestive signals & processes.

ADRENAL MEDULLA; neurohormones *adrenaline* & NA.

STEROID HORMONES; like thyroxines, are small, and bind other zinc-finger proteins, with pervasive effects, and also act via cell wall receptors, like most other hormones.

ADRENAL CORTEX; *catabolic* steroids (e.g. *cortisol*) inhibit gene expression, others rule salt-water balance. Makes extra sex steroids, 'male' and 'female', in both sexes.

OVARIES; womanhood via *estrogens* and *progesterones*.

TESTES; manhood via androgens like *testosterones*.

PLACENTA; *steroids* and enormous *pregnancy peptides*.

ALL OTHER ORGANS; make hormone like substances.

Timeline of Life on Earth

billions of years ago

13.7	Big Bang
13.3	First stars
13.1	Earliest galaxies and galaxy clusters,
10	First metal-rich stars
10 – 6.5	Formation of Milky Way.
4.8	Nearby supernova triggers formation of Sun.
4.7	Jupiter, Saturn, Uranus & Neptune form.
4.6	Sun ignites hydrogen fusion
4.5	Earth, Moon, Mars, Venus & Mercury formed.
4.4	Water delivered to Earth.
4.1	Earth's crust solidifies. Atmosphere & oceans form.
4	Oldest rocks on Earth.
3.9	Late Heavy Bombardment of meteoroids.
3.8	First simple cells (chemoautotrophs).
3.5	Last universal common ancestor.
3 – 2.5	First photosynthesis in cyanobacteria. Atmospheric oxygen rises too high and poisons most other life.
2	First eukaryotic cells in oceans.
1.2	Simple multicellular beasties in oceans. Sex.

millions of years ago

580 – 540	Atmospheric oxygen creates ozone layer permitting colonisation of the land.
570 – 500	Cambrian Period. Trilobites, arthropods.
560	First fungi, sponges, ctenophora (comb-jellies), jellyfish, coral & anenomes.
530	Lampreys, protostomes & deuterstomes (e.g. flatworms, velvet worms, molluscs) & sea squirts. First known footprints on land.
500 – 440	Ordovician Period. First cephalopods (nautiloids).
485	First vertebrates with true bones (jawless fish).
475 – 430	First land plants (liverworts) and land fungi .
445	Ordovician-Silurian extinction event. 60% of marine invertebrates die due to global cooling.
440	Ray-finned fish (e.g. herring, salmon, sturgeon).
440 – 410	Silurian Period. Sea & land scorpions & lungfish.
410 – 360	Devonian Period. Insects, teeth, seeds, spore-trees.
375 – 360	Late Devonian extinction event. 70% of all species disappear due to ocean volcanism and comets.
360 – 320	Carboniferous Mississippian Period. Crabs, ferns, brachiopods and amphibians (frogs & salamanders). First gymnosperms (conifers and ginkgos).
320 – 290	Carboniferous Pennsylvanian Period. Giant insects.
290 – 250	Permian Period. First beetles, reptiles, turtles and crocodilians (245 MYA).
251	Permian-Triassic extinction event. Environmental changes and other catastrophic events. 96% of marine and 70% of land species are wiped out.
250 – 205	Triassic Period. First dinosaurs. Snakes (220 MYA).
215	First mammals.
205	Triassic-Jurassic extinction event caused by volcanos or asteroid(s). 50% of all species vanish.
205 – 140	Jurassic Period. Quarry Dinosaurs, duck-billed Platypus (180 MYA), rays and birds.
130	First flowers and flowering trees (angiosperms).
140 – 65	Cretaceous Period. Marsupials (kangaroos, opossums) (140 MYA).
105 – 80	All other placental mammals (elephants, manatees, aardvarks). First bees, ants, and termites.
85	Laurasian continent descendents (e.g. cats, dogs, camels, horses, seals, whales, hippos, bats).
75	Rodents & rabbits (com. ancestor around 40 MYA).
70	Tree shrews etc.
65.5	Cretacious-Tertiary extinction event caused by volcanos or asteroids. 75% of species vanish.
65 – 1.5	Tertiary Period. Conifers and Ginkos.
40-14	Butterflies and moths, deer, grasses, giraffes, capuchins, marmosets, spider monkeys, macaques, colobus, baboons, gibbons and then orangutans.
7 – 6	First humanish species (Sahelanthropus tchadensis). Gorillas (7 MYA). Orrorin tugenesis (6 MYA).
5.8 – 4.4	Mammoths, tree sloths and hippopotami.
4 – 3	Australopithicus anamensis (4 MYA); Australopithicus afarensis (3.9 MYA); Australopithicus africanus (3 MYA)
2	Chimpanzee & Bonobo diverge from com. ancestor.
2.4 – 0.2	Homo habilis (2.5 - 1.9 MYA); Homo rudolfensis (2.4 MYA); Homo ergaster (1.9 - 1.5 MYA); Homo erectus (1.4 - 0.2 MYA).

thousands of years ago

800 – 300	Homo heidelbergensis; Homo neanderthalensis (300 TYA); Homo sapiens (300 TYA).
30	Neanderthals extinct.
2 – now	Sixth mass extinction event, caused by humans.
future	anyone's guess ...

THREEFOLD SYSTEMS

	ECTODERM	MESODERM	ENDODERM
Embryonic germ cell layer	thin, wiry, active	stocky, strong, cyclic	rounded, steady, slow
Body type			
Ayurvedic dosha & humour	VATA - WIND	PITTA - BILE	KAPHA - PHLEGM
Tibetan elements	AIR & SPACE	FIRE	EARTH & WATER
Arabian-European alchemy	mercury/communication	sulfur/energy	sal/structure
Fluid compartment & cation	extracellular/calcium	blood/sodium	intracellular/potassium
Body systems	neuro-endocrine, integration	metabolic, immune	assimilation, excretion
Tissue & organ colour	brain nerve skin white/brown	muscle liver heart red	gut lung tubes transparent
Teeth	long crooked brown-black	medium hard grey-yellow	strong large white-blue
Eyes	beady, black holes	penetrating, yellow-red	big, open, clear
Psyche	thought, integration, desire	compassion, vision, anger	acceptance, stability, greed
Life	inspirational, myriad, future	phasic, powerful, present	devoted, belonging, past
Memory	quick/recent	all encompassing	slow, long, deep
Sleep & dreams	shallow, mountains, flying	refreshing, action, light	sound, water, clouds
Trouble	cold, falling, isolation	heat, violence, gloom	drowning, stasis, loss

WESTERN FOURFOLD SYSTEM

Type	ARTISAN	GUARDIAN	IDEALIST	RATIONAL	Plato
Element	FIRE	EARTH	AIR	WATER	Nature
Qualities	hot & dry	cold & dry	hot & wet	cold & wet	Weather
Solid	tetrahedron	cube	octahedron	icosahedron	Geometry
Humour	SANGUINE	MELANCHOLIC	CHOLERIC	PHLEGMATIC	Galen
Fluid & organ	blood & liver	bile & intestines	plasma & lungs	mucus & kidneys	Medieval
Extremity	arms	legs	head	genitals	Body politic
Systems	electric-metabolic	structural-permanent	breath-integration	reproduction-homeostasis	20th C.
Temperament	changeable	industrious	inspired	curious	Paracelsus
Personality	influential	conscientious	dominant	steady	Haines
Jungian types	ISFP, ISTP,	ISFJ, ISTJ,	INFJ, INFP,	INTJ, INTP,	Jung
(see below)	ESFP, ESTP	ESFJ, ESTJ	ENFJ, ENFP	ENTJ, ENTP	
Nature	exploitative	hoarding	receptive	marketing	Fromm
Mindedness	probing	scheduling	friendly	tough	Myers-Briggs
Self Image	artistic, audacious	dependable, beneficent	empathic, benevolent	ingenious, autonomous	Keirsey
	adaptable, hedonistic	respectable, stoical	authentic, altruistic	resolute, pragmatic	
Orientation	optimistic, cynical	pessimistic, fatalistic	credulous, mystical	skeptical, relativistic	
	here, now	gateways, yesterday	pathways, tomorrow	intersections, intervals	

Jung's categories are based on four axes: Introverted or Extroverted (I or E), Intuitive or Sensing (N or S), Feeling or Thinking (F or T), and Perceiving or Judging (P or J). See too page 271.

CHINESE FIVEFOLD SYSTEM

	木 WOOD	火 FIRE	土 EARTH	金 METAL	水 WATER
Yin Organ (zang)	liver	heart & pericardium	spleen	lung	kidney
in charge of	strategy, plans	pulse & protection	processing	qi of heaven	water control
Muscle meridians	leg jue yin	arm shao & jue yin	leg tai yin	arm tai yin	leg shao yin
Yang Organ (fu)	gall bladder	sm. intest & trip heat	stomach	large intestine	bladder
in charge of	decisions	digestion & regulation	fermentation	elimination	storing liquids
Muscle meridians	leg shao yang	arm tai & shao yang	leg yang ming	arm yang ming	leg tai yang
Tissue	tendons, sinews	blood vessels	flesh, muscles	skin, pores	bones, marrow
reflects in	nails	face, complexion	lips	body hair	head hair, teeth
Orifice, sense	eyes, sight	tongue, speech	mouth, taste	nose, smell	ears, hearing
Body fluid	tears	sweat	drool	mucus	spittle
Taste, odour	sour, rancid	bitter, scorched	sweet, aromatic	pungent, rotten	salty, putrid
Voice, emotion	shouting, anger	laughing, elation	singing, worry	crying, sadness	groaning, fear
Aspects of Shen	*hun*, soul life	*shen*, awareness	*yi*, memory intellect	*po*, instinct	*zhi*, will power
Stage	germination	growth	transformation	harvest	storage
Season	spring	summer	late summer	autumn	winter
Injurious climate	windy	hot	damp, humid	dry	cold
Direction	east	south	center, middle	west	north
Colour, time	green, dawn	red, midday	yellow, afternoon	white, dusk	black, midnight
Animals	dragon, sheep	phoenix, fowl	pangu, snake, ox	tiger, dog	tortoise, pig
Planet	jupiter ♃	mars ♂	saturn ♄	venus ♀	mercury ☿
Food	wheat, lemons	pepper, alfalfa, greens	millet, potato, fruit	rice, ginger, air	beans, seafoods
Yin-yang	lesser yang	utmost yang	center	lesser yin	utmost yin
Mode, note	*jiao*, e	*zhi*, g	*gong*, c	*shang*, d	*yu*, a

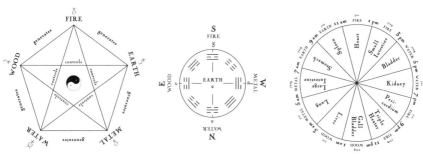

Northern sky

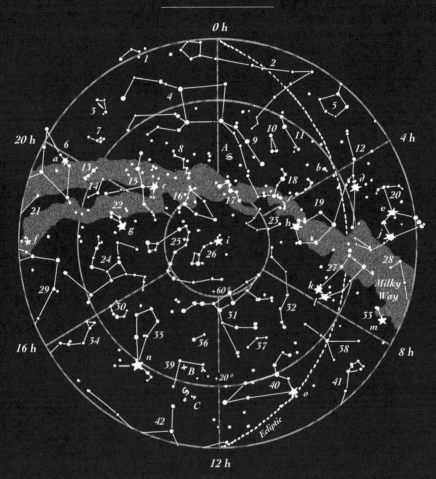

0 h
20 h
4 h
16 h
12 h
8 h

A
a
b
c
d
e
f
g
h
i
j
k
l
m
n
o

+60°
+20°

Milky Way

Ecliptic

STARS: *a. Altair, b. Pleiades, c. Deneb, d. Aldebaran, e. Bellatrix, f. Barnard's Star, g. Vega, h. Capella, i. Polaris, j. Betelgeuse, k. Castor, l. Pollux, m. Procyon, n. Arcturus, o. Regulus. Misc: A. Andromeda galaxy, B. North galactic pole, C. Coma and Virgo galaxy clusters. CONSTELLATIONS: 1. Aquarius, 2. Pisces, 3. Equuleus, 4. Pegasus, 5. Cetus, 6. Aquila, 7. Delphinus, 8. Lacerta 9. Andromeda, 10. Triangulum, 11. Aries, 12. Taurus, 13. Sagitta, 14. Vulpecula, 15. Cygnus, 16. Cepheus, 17. Cassiopea, 18. Perseus, 19. Auriga, 20. Orion, 21. Serpens Cauda 22. Lyra, 23. Camelopardalis, 24. Hercules, 25. Draco, 26. Ursa Minor, 27. Gemini, 28. Monoceros, 29. Ophiuchus, 30.Corona Borealis, 31. Ursa Major, 32. Lynx, 33. Canis Minor, 34. Serpens Caput, 35. Bootes, 36. Canes Venatici, 37. Leo Minor, 38. Cancer, 39. Coma Berenices, 40. Leo, 41. Hydra, 42. Virgo*

Southern Sky

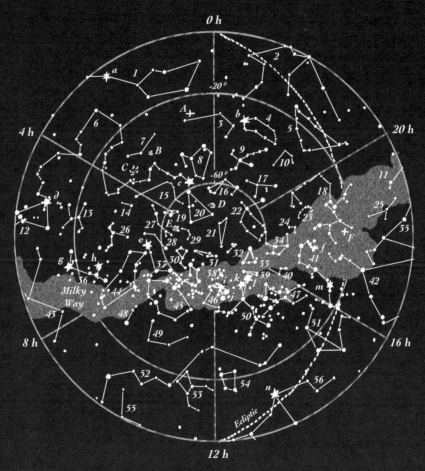

STARS: *a. Mira, b. Formalhaut, c. Achernar, d. Rigel, e. Canopus, f. Shaula, g. Sirius, h. Adhara, i. Acrus, j. Mimosa, k. Hadar, l. Rigel Kentaurus, m. Antares, n. Spica. Misc: A. South galactic pole, B. Fornax galaxy sys. C. Fornax galaxy clus. D. Small Magellanic cloud. E. Large Magellanic cloud. F. Galactic center CONSTELLATIONS: 1. Cetus, 2. Aquarius, 3. Sculptor, 4. Piscis Austrinus, 5. Capricornus, 6. Eridanus, 7. Fornax, 8. Phoenix 9. Grus, 10. Microscopium, 11. Aquila, 12. Orion, 13. Lepus, 14. Cælum, 15. Horologium, 16. Tucana, 17. Indus, 18. Sagittarius, 19. Reticulum, 20. Hydrus, 21. Octans, 22. Pavo, 23. Corona Australis, 24. Telescopium, 25. Scutum, 26. Columba, 27. Dorado, 28. Pictor, 29. Mensa, 30. Volans, 31. Chameleon, 32. Apus, 33. Triangulum Australæ, 34. Ara, 35. Serpens Cauda, 36. Canis Major, 37. Carina, 38. Musca, 39. Circinus, 40. Norma, 41. Scorpius, 42. Ophiuchus, 43. Monoceros, 44. Puppis, 45. Vela, 46. Crux, 47. Lupus, 48. Pyxis, 49. Antilia, 50. Centaurus, 51. Libra, 52. Hydra, 53. Crater, 54. Corvus, 55. Sextans, 56. Virgo*

GALACTIC MAPS
superclusters, the local supercluster, and the local group

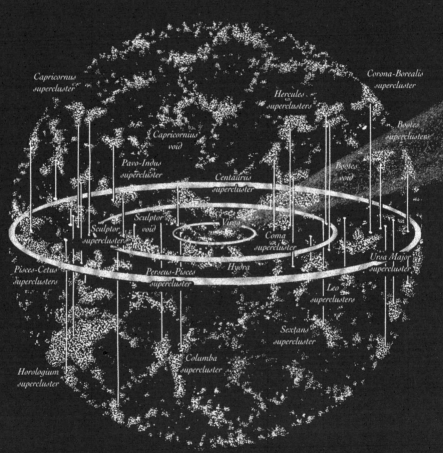

Capricornus
supercluster

Corona-Borealis
supercluster

Hercules
superclusters

Boötes
superclusters

Capricornus
void

Pavo-Indus
supercluster

Boötes
void

Centaurus
supercluster

Sculptor
void

Virgo

Coma
supercluster

Sculptor
supercluster

Pisces-Cetus
superclusters

Perseus-Pisces
supercluster

Hydra

Ursa Major
supercluster

Leo
superclusters

Sextans
supercluster

Columba
supercluster

Horologium
supercluster

Superclusters
rings are spaced
400 million light-years apart

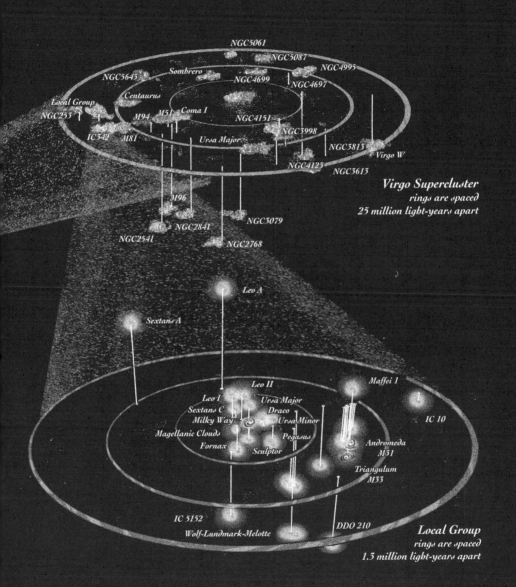

NGC5061

NGC5087

NGC4995

NGC5643

Sombrero

NGC4699

NGC4697

Local Group

Centaurus

Coma I

NGC4151

NGC253

M94

M51

NGC3998

M81

Ursa Major

NGC3813

IC342

M81

Virgo W

NGC4125

NGC3613

M96

Virgo Supercluster
rings are spaced
25 million light-years apart

NGC3079

NGC2841

NGC2541

NGC2768

Leo A

Sextans A

Leo II

Maffei 1

Leo I

Ursa Major

Sextans C

Draco

IC 10

Milky Way

Ursa Minor

Magellanic Clouds

Pegasus

Fornax

Sculptor

Andromeda
M31

Triangulum
M33

IC 5152

Wolf-Lundmark-Melotte

DDO 210

Local Group
rings are spaced
1.3 million light-years apart

COMMON ROCKS AND MINERALS

MINERALS *are naturally occurring solid chemical substances formed through geological processes. Two classes of minerals are generally recognized: MAFIC MINERALS and FELSIC MINERALS. The nine minerals listed below constitute 95% of the Earth's crust.*

MAFIC MINERALS *are normally dark in colour and enriched in the heavier elements of Fe and Mg. Principle mafic minerals follow (the first four being very dark to greenish and the last light to transparent):*

BIOTITE is a magnesium enriched mica mineral. Structurally a sheet silicate whose sheets are weakly bonded by potassium ions. General formula $K(Mg,Fe)_3AlSi_3O_{10}(F,OH)_2$. **AMPHIBOLE/ HORNBLENDE** is a dark silicate with double chain SiO_4 tetrahedra laced with Fe and Mg, and sometimes Ca; With General formula $(Ca,Na)_{2-3}(Mg,Fe,Al)_5(Al,Si)_8O_{22}(OH,F)_2$. **PYROXENES/AUGITE** comprises a group of single chain silica monoclinic and orthorhombic tetrahedra. $XY(Si,Al)_2O_6$, where X can be Ca, Na, Fe, Mg, Zn, Mn & Li and Y can be Cr, Al Sc etc. **OLIVINE** is a magnesium iron silicate general formula $(Mg,Fe)_2SiO_4$, one of the most common minerals found on Earth and in meteorites. **Ca-PLAGIOCLASE** is a tectosilicate mineral within the feldspar group which comprises approximately 60% of the Earth's crust; $CaAl_2Si_2O_8$.

FELSIC MINERALS *are normally light coloured and enriched in the lighter elements, such as Si, O2, Al and K. The principle felsics:*

QUARTZ is a continuous tetrahedral crystal SiO_2 and the second most abundant mineral in Earth's crust after feldspar. **MUSCOVITE** is a transparent K enriched mica mineral; with general formula $K_2Al_4(Si_6Al_2)O_{20}(OH,F)_4$. **ORTHOCLASE**: is a K enriched feldspar; $KAlSi_3O_8$. **Na-PLAGIOCLASE** is another member of the feldspar family; $NaAlSi_3O_8$.

ROCKS *are aggregates of minerals and come in three basic classes.*

IGNEOUS ROCKS *form through the solidification of molten rocks. INTRUSIVE IGNEOUS ROCKS [IIR] solidify under the Earth, while EXTRUSIVE IGNEOUS ROCKS [EIR] solidify on or above the surface. Here are the most common igneous rocks:*

ANDESITE is a fine-grained grey EIR, mainly plagioclase with hornblende, pyroxene and biotite. **BASALT** is a fine-grained, blackish EIR, mainly plagioclase and pyroxene. **DIORITE** is a coarse-grained, IIR mixture of feldspar, pyroxene, hornblende and sometimes quartz. **GABBRO** is a coarse-grained, dark IIR containing feldspar, augite and occasionally olivine. **GRANITE** is a coarse-grained, lightish, IIR, mainly quartz and feldspar, which can take many colours. **OBSIDIAN** is a dark volcanic glass formed from such quick rock cooling that crystals do not form. **PERIDOTITE** is a coarse-grained IIR composed almost entirely of olivine. **PUMICE** is a pale lightweight vesicular igneous rock, a solidified larval froth, normally silicic or felsic. **RHYOLITE** is a pale, fine-grained EIR that contains quartz and feldspar. **SCORIA** is a dark vesicular EIR, formed as a frothy crust on the top of a larva flow, normally basaltic or andesitic.

METAMORPHIC ROCKS *are created by heat, pressure and chemistry. There are two categories: FOLIATED METAMORPHIC ROCKS [FMR] with a banded/ layered appearance, and NON-FOLIATED METAMORPHIC ROCKS [NMR] are characteristically devoid of any layering or internal structure. These are the most common metamorphic rocks:*

AMPHIBOLITE is a NMR composed mainly of amphibole and Plagioclase recrystallized under high temperature and pressure. **GNEISS** is a FMR with a granular and banded appearance containing quartz and feldspars. **HORNFELS** is a fine-grained baked NMR with no specific composition. **MARBLE** is a pale NMR produced from the metamorphism of limestone, so mostly comprised of calcium carbonate. **PHYLLITE** is a sometimes lustrous and wrinkled FMR made of very fine mica, halfway between slate and schist. **QUARTZITE** is a pale FMR of mainly quartz, formed through metamorphism of sandstone. **SCHIST** is a FMR with high foliation and large amounts of mica. **SLATE** is a grey FMR formed through the metamorphism of shale.

SEDIMENTARY ROCKS *form by the accumulation of sediments. CLASTIC SEDIMENTARY ROCKS [CLSR] are formed from the debris of mechanical weathering. CHEMICAL SEDIMENTARY ROCKS [CHSR] are formed when dissolved minerals precipitate from solution. BIOCHEMICAL SEDIMENTARY ROCKS [BSR] form from the accumulation of plant or animal debris. A fourth category is formed by impacts and volcanism. Here are the most common ones:*

BRECCIA is any CLSR composed of large angular fragments cemented by a fine-grain matrix. **CHERT** is a microcrystalline sedimentary rock composed mostly of SiO_2. Chalcedony and flint are both chert. **COAL** is an OSR formed mainly from plant debris. **CONGLOMERATE** is any CLSR of large rounded particles, smaller particles, and cement. **IRON ORE** e.g. hematite, is a CHSR formed when Fe and O_2 combine in solution and deposit as a sediment. **LIMESTONE** is primarily calcium carbonate and can be a CHSR, precipitating from water, or an BSR, forming from shells, corals and algal debris. **ROCK SALT** is a CHSR formed by evaporation. **SANDSTONE** is a CLSR made from compressed sand (mainly SiO_2). **SHALE** is a CLSR made of clay-sized (less than 1/256mm) debris, mostly mud (soil, silt and clay) and tiny mineral fragments

GLOSSARY INDEX

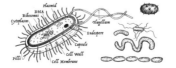

Rhizobium spores infect root hair

The Nodule begins to develop

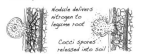

Nodule delivers nitrogen to legume root

Cocci spores released into soil

405

Fusion, nuclear 160; stellar 342-343

Gaia 252-253

Galapagos archipelago 188-189

Galaxies 328; early 324; types 330-332

Galaxy clusters 324, 328; local cluster 401

Galileo, 94

Gamete 200, 272; haploid inter-generational cells. Sperm or egg (*below*).

Ganglia 300; nodal colonies of nerve cells in the brain and spinal cord, or just outside the spine (which has two sympathetic and two sensory ganglia per segment).

Gases, 134; pressure in 98-99

Gating 268; central nervous blocking effect by a pattern of peripheral stimuli.

Gauss, summing theorem 40; lens equation 102; Gaussian distribution 84;

Gears, equations for 92

Gene 190-193, 208-209, 272, 274, 276, 278, 280, 282, 286, 394; a stretch of DNA which codes for a protein (or RNA unit); humans have about 30,000 genes but can make ten times as many proteins, as RNA combines recipes.

Gene Pool. The complete set of unique alleles in a species.

General Relativity 115

Genetic engineering 254-255

Genome 194-195, 274; total DNA for a species.

Genotype. Describes the genetic constitution of an organism.

Geology 354-357, 402

Germanium 144

Glial cells 300; metabolic and structural support for nerves in brain and spinal cord.

Globular clusters 338-339

Glomeruli 290; initial filtration subunits of the kidney; each consists of a high pressure capillary bundle and collecting capsule.

Glucose 134-135, 232, 236; molecule 393

Gluons 168, 170-171

Goldbach conjecture 8

Golgi Apparatus 237

Gonads 282, 285; ovaries or testes.

Gradients 66-67. The gradient of a line is a ± measurement of its slope, found by dividing the vertical by the horizontal displacement. The gradient of a horizontal line is zero, and the gradient of a vertical line is not defined.

Gradualism in evolution 204

Granulocytes 298; white blood cells, mostly neutrophils which deal with bacteria.

Granum, in chloroplast 237

Graph. The graph of the function f plots the points (x, f (x)) on a coordinate plane.

Graphite and graphene 146-147

Gravitation, equations 88; constant 88; force 168-169, 362-363; lensing 362-363

Great Walls 325-325, 328, 364

Growth hormone 308, 312, 394; pituitary peptide, rules regeneration and integration.

Guppy Fish 216

Gut 201, 224, 244, 246, 280, 283, 288-289, 296, 392

Habitable zones 374-375

Hadrons 172-173, 322-323, 385

Hæckel, Ernst 185

Hæmoglobin 290, 294; gas-carrying molecule.

Half-life 160-1

Halogens and halides 154-155

Hamiltonian operators 116-117

Haploid 273; single set of chromosomes.

Heart 292-3

Heisenberg uncertainty principle 116

Heliosheath and heliopause 348

Helium 140, 322

Hertzsprung-Russell diagram 341

Holographic universe 378-381

Homeostasis 252-253, 310-311; keeping the balance.

Homo 302; Latin for 'man', as in H. sapiens.

Hooke's Law 96

Hormones 270, 282-285, 308-312, 394-395; the most influential signalling molecules in your body.

Hubble's Law 358

Human being, consistency of 122

Humours 270-271

Hydrocarbons 152-153, 388

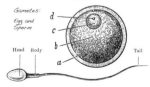

Gametes: Egg and Sperm

Head Body Tail

Hydrogen, bonds 135-137, 140, 150; in body 274; Loose proton links between molecules, especially important in life-temperature aqueous systems (like you).

Hyperspheres 118

Icosahedron 47

Identity. Another name for an equation.

Immunology, human 298-299

Immunication 312, 393; the many ways in which immune system cells communicate.

In Utero 286; while in your mother's womb.

Inductors 105. Coils which convert electric current into magnetism.

Inflammation 312; it hurts because tissues are being opened up to allow repair.

Insulin 308, 394; anabolic pancreatic hormone.

Integration 110-111

Intestines 288-289

Intracellular fluid 279, 280, 392; 'cytosol'.

Introns 194, 275; palindromic or unwanted regions of mRNA which are snipped out before protein synthesis.

Inverse trigonometric functions arcsin, arccos, arctan (also written *** sin-1, cos-1, tan-1) are such that arcsin(x) = q, where sin q = x, etc.

Invertebrates 240-241

Ionic bonds 136

Ions 136, 150-151, 278, 300, 306, 310, 392; charged atom(s), eg Na+, Cl-, PO43-, usually in water.

Iris, of eye 266

Iron 148, 156-157, 255, 294, 342, 393

Isotopes 130, 140-141

Jellyfish 212, 240, 242, 395

Junk DNA 194

Kelvin scale 98

Kepler's laws of planetary motion 86

Kidneys 280, 282, 284, 290, 310, 396.

Kin Kindness 214-215

Kingdoms, classification of 184, 224-5

Knots, polygons from 36-37; in 3D 118

Kreps cycle 393

Kuiper belt 348-349

Kundalini 268; awakened earth energy.

Lamarck, Jean-Baptiste 184, 186, 202

Lamarckism. A theory that some acquired

Adenine (A)
Dna / rna

Cytosine (C)
Dna / rna

Thymine (T)
Dna

Status quo 310; dynamic equilibrium of existing conditions, current state of play.

Steady state cosmology 370-371

Stellar nurseries 346

Stoma 237; plant pores for gas exchange.

Stomach, human 289

Stomata 237; plant pores, may close when gaseous exchange is unnecessary.

Strain 96-97

Stress 96-97

String Theory 176-177, 362-363

Stroma, 237; the fluid between grana

Strong nuclear force 166, 168-169, 362-363

Strontium 142

Sulfur 122, 138, 148-149, 152, 224, 228, 248, 253, 392

Summing, square and triangular numbers 38-39; rectangular and cubic numbers 40-41

Sun 86, 140, 350-351

Superclusters 259, 324, 328; local 400

Supercoiling 274; keep twisting a rubber band to make a simple superhelix.

Supergiant 340-341; very big star

Supernovæ 324, 340-341, 342, 344, 352, 360, 362, 372, 374, 395

Supersymmetries 176-177

Surface areas 64-65

Symbiosis 210-213

Sympathetic; division of autonomic nervous system, running high-alert catabolic phases of action eg hunting and lovemaking.

Synapse 300; joint between two nerve cells; a neuromuscular junction is a sort of synapse.

Tan half formulæ 387

Tangents in a triangle 68; trigonometrical identities 70: Law of 72

Taylor Expansion 387

Telomeres 221; at tips of chromosomes

Tension 97. Tension and its opposite, compression, are measured as forces.

Tensors 115; matrices used in relativity

Thymus 298, 309, 394; immune-endocrine gland.

Thyroxins 308, 394; hormones secreted by the thyroid; major metabolic influence.

Time travel 368-369

Tissue 266, 278, 280-282, 284, 286, 294, 296, 393.

Titanium 156

Torque 92

Tortoise 188-189

Transgenic rice 255

Transistors 108-109

Transition metals 156-157

Trapezium. A quadrilateral with one pair of opposite sides parallel.

Triangle, area of 4; angles 12-13; formulæ 60-61

Trigonometry 68-69; identities 70-71; spherical 72-73

Triploid 193

Tritium 140-141

Trophic levels 248-249

Twin slit experiment 174-175

Uncertainty principle 116, 174-175

Unit. A unit is a standardized amount of a measurable quantity. Units must be used consistently, e.g. if acceleration is taken in [meters/sec]/sec, all distances covered must be measured in meters and all durations in seconds.

Units, table of 384

Universe 140, 148, 150, 166, 168, 174, 176, 258, 268, 318, 322-325, 326, 332, 334, 366, 372; plasma 364-365; open, flat or closed 370-371, fine tuning in 376-377; other models of 378-379

Uranium 158, 160

Uterine milk 282; nutritious early food.

Vacuole 237; membrane-bound water-filled organelle found in all plant and fungal cells.

Valence, electrons 134

Vampire bats 214-215

Van der Waals forces 136

Variable, a symbol representing a varying or unknown quantity.

Vasopressin 394; blood pressure, memory, and salt balance hormone.

Vectors 76; movements in the plane.

Veins, in leaf 237; in human body 294-295

Velocity. The distance an object travels per unit time, in other words its speed, in a particular direction.

Ventricles 292; two lower chambers of the heart; pump blood out to body and lungs.

Vertebrates 218, 240-241, 244, 395

Vertex, vertices 50-51

Villi 288, 290; intestinal fingers, with microvilli, central capillaries and a lymphatic, which absorb all your food from the small intestine.

Virgo 398-399; cluster 328, 401

Viruses 212, 227, 274; parasitic polyhedral replicators.

Voids 326-329, 372

Voltage 104-105. An 'electromotive' force which causes a flow of current.

Voltage gate 300; door of a channel in cell wall that opens or shuts depending on local voltage.

Voltage gradient 300; change in electrical charge per unit distance.

Volumes 64-65, prism and cylinder 16

Waddington, Conrad Hal 202

Water 135, 137, 150-151, 393

Waves 100-101. A wave is a vibration propagated through a medium which involves a periodic interchange of a pair of quantities. Its period is related to its wavelength.

Weak nuclear force 168-169, 362 363

Weight. The weight of an object is the force produced by its mass undergoing gravitational acceleration. We experience mass as weight.

Weights and measures, conversions 386

Wizzard 58, 69-119, 122, 379

White Dwarfs 340-341

White matter 302; inner parts of brain and outer parts of spinal cord; mostly consists of axons, getting its colour from their myelin.

Work 90

Worms 183, 240-241, 244-245, 246, 395

Wormhole 334, *below*

X-Rays 360-361

Y-Chromosome 208-209, 282

Yang 268, 396; positive, male, light, active.

Yin 268, 396; negative, female, dark, receptive.

Yolk sac 282; embryonic food and cell store.

Young's Modulus, in a material 96

Z the end.

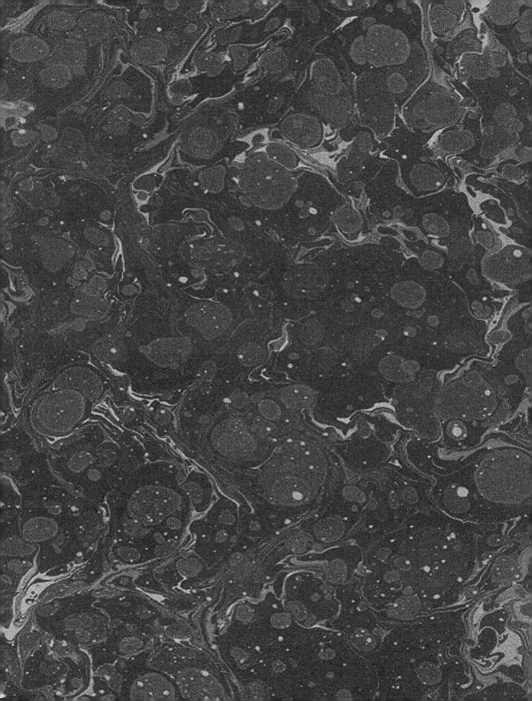